AF824566

GROVE'S FIELD GUIDE TO CLIMATE CHANGE

Prospering in a Post-Climate Change World

Bob Grove

Wordsinger Publishing

ISBN: 978-1-961014-08-4
LCCN: 2023906009

Published in the United States of America

Wordsinger Publishing
Hilo, Hawaii
www.Wordsinger.com

CONTENTS

INTRODUCTION

Face it. Climate change has happened. It's no longer a future threat. We're living with it today. Let's just accept that and move forward. We have to survive an expanding range of weather extremes, an ever growing list of natural disasters, a steady drip of pandemic-prone viruses, and the economic impacts of skyrocketing energy and food costs. The environment is filled with new challenges but we can survive; even prosper.

The science behind climate change is pretty simple if we ignore the politics. When we say climate, we're typically talking about the 30-year average of a region's long-term weather patterns; the average of things like temperature and rainfall. When a region experiences climate change, it just means the area's long-term weather patterns are moving outside their normal range. The area has become too hot or too cold; too wet or too dry.

We say "too" because local plants, animals and infrastructure are usually adapted to an area's climate. And when the climate changes too much, plants and animals may die and infrastructure may collapse. People that are dependent on these food systems and infrastructure may also die or suffer serious economic damage. Climate change is usually accompanied by weather disasters. Heat waves, blizzards, droughts, floods, wildfires, tornadoes and hurricanes, are all natural weather disasters that result from various combinations of too hot, too cold, too wet or too dry – that is, climate change.

When weather disasters start happening more often, it's a pretty clear indication of severe climate change. The smart people at NAA (National Oceanic and Atmospheric Administration) report[1] that during the 1980s, the average number of U.S. weather disasters causing over one billion dollars in damage was a little over three per year. The average during the 1990s was almost six per year. The average during the 2000s was seven and the average in the 2010s leaped to 13 weather disasters per year. If we look at a more recent 5-year average (2018-2022), it's 18 disasters per year. And an even more recent 3-year average (2020-2022) is 20 per year. Do you see a trend?

Average Number of $Billion Weather Disasters per Year

1980s – 3
1990s – 6
2000s – 7
2010s – 13
2020s – 20

The growing number of billion-dollar weather disasters – not to mention all the measly little multi-million dollar disasters that are too small for the government to count – are a pretty clear indication of how severe climate change has become. The big weather disasters are now happening three times more often then average and all indications are that weather disasters will continue to increase in size and frequency.

If weather disasters are associated with the climate changing, what causes the climate to change? Climates are defined as a region's long-term average temperature range and rainfall.[2] If a region's long-term average temperature goes up or down, then we're going to see a comparable change in that region's climate.

Scientists started recording atmospheric temperatures around the world in 1880 and now have well over a century of global thermometer readings to study. They also have ice cores from Greenland and Antarctica that contain ice layers over 800,000 years old. Ice cores, like tree rings, are composed of discernable annual layers, and scientists are smart enough to measure the types of oxygen in each layer[3] to determine the average temperature for that year. So scientists have a solid 800,000 years of average annual temperatures to compare. They report that the planet has gone through several ice ages and, after each ice age, the planet warms an average of 5.5°C over the following 5000 years. That's an average rise of .0011°C per year. In comparison, the average annual temperature over the last 100 years his risen .007°C per year, and supercomputer weather models predict that the temperature over the next century will increase at 20 times that rate.[4]

Keep in mind that all climate temperature stuff is discussed in Celsius, the metric system used everywhere except in Myanmar, Liberia, and the United States (yeah, I know). And one degree Celsius is BIGGER than one degree Fahrenheit – 1.8% bigger. So when you hear that the average temperature of the planet is going up 1.5°C (Celsius) and you think that's nothing, remember that it's actually 2.7°F (Fahrenheit) in U.S. speak. Okay, okay ... I hear you saying that it's still nothing. But given the tremendous size of the planet's oceans and atmosphere, it takes a massive amount of heat energy to raise the global temperature even a fraction of a degree. Consider that the average human body temperature is 98.6°F and that a 2.7°F increase in our body temperature would mean we're running a fever of 101.3°F. Most sources recommend seeing a doctor if your temperature is over 100.4°F for more than three days. And they say you should head for the emergency room if your temperature hits 104°F. If you want to move that back into climate speak, the planet has hit emergency room status when it's average surface temperature has increased 3°C.

You can see why United Nations agreements like the 2015 Paris Climate Accord[5] and the 2021 Glasgow Climate Pact[6] make such a big deal about keeping the planet's global warming fever below 1.5°C. But the planet's temperature is already up 1.6°F (.89°C)[7] and the World Meteorological Organization's Annual Climate Update for 2022 notes that the likelihood of exceeding 1.5°C, has gone from 10% in 2017-2021 to 48% in 2022-2026.[8]

Some NGO (Non-Governmental Organization) sources, like Climate Action Tracker,[9] are already calculating a 2.9°C temperature increase over the next century based on actual governmental policies vs. what they've pledged. They point out that the gap between governmental policies and pledges is the equivalent of 19-20 gigatons of carbon equivalents ($GtCO_2e$) annually. Numbers vary but the clear consensus is that we need to prepare for significantly warmer temperatures worldwide – that is, global warming.

Which brings us to the next question. If the planet's average surface temperature is warming and causing climate changes that increase the number and severity of weather disasters, why is the planet getting warmer?

It's worth mentioning that variations in the earth's orbit caused by the gravity of big planets like Jupiter, change the earth's distance from the sun over a long 80,000-100,000 year cycle. And it makes sense that if we're further from the sun, it's going to get cooler, and if we're closer, things will get hotter. But scientists tell us that orbital variations aren't big enough to create ice ages. Besides, our orbit is actually moving us away from sun right now and, by this logic, we should be heading into an ice age, not getting warmer. So what's with that? Turns our that atmospheric carbon dioxide concentrations play a more significant role than orbital variation in the cycle of ice ages.[10] Ice cores show that the atmospheric concentration of carbon dioxide over the past 800,000 years has consistently cycled from about 180 parts per million (ppm) to 280 ppm in step with the coming and going of ice ages. But over the past 150+ years, carbon dioxide concentrations have gone from 280 ppm to over 420 ppm, a change that lines up perfectly with the current rate of global warming.[11] It's the concentration of carbon in the atmosphere that is the primary cause of global warming. Hence the earlier comment about the difference between government policies and pledges being 19-20 gigatons of carbon equivalents annually. We can control global warming by controlling the concentration of carbon in our atmosphere.

Carbon, and its equivalents, have the unique ability to absorb light reflected off the planet's surface, and convert it to vibrational energy, which then heats up the atmosphere.[12] It's the same process that makes a greenhouse toasty warm inside when it's snowing outside. Hence the adoption of the phrase "greenhouse gas" (GHG) to describe carbon dioxide and its equivalents.

You might wonder why greenhouse gases absorb light reflected off the planet instead of grabbing it as it comes in. I know I did. Turns out that different gases absorb different wavelengths of light, due to their different atomic structures. The light coming from the sun tends to be shorter wavelengths in the visible light spectrum, while the light bouncing off the planet tends to be longer wavelengths in the infrared light spectrum. You've probably seen those infrared "heating" lamps keeping the food on your favorite buffet warm. Turns out that greenhouse gases are best at absorbing the long infrared wavelengths and not letting them leave the atmosphere. So our planet's surface is warming up just like that buffet. The EPA identifies four major gases that are responsible for global warming.[13] Carbon dioxide, methane, nitrous oxide, and fluorinated gases – that is, greenhouse gases.

Carbon dioxide (CO_2) is 79% of greenhouse gas emissions in the United States. It results from burning things like coal, oil, natural gas, solid wastes, trees and other biological materials; and from certain chemical reactions, such as manufacturing cement. We've already talked about CO_2 swinging from 180 ppm to 250 ppm over the last 800,000 years of ice age cycles. And we've talked about how it has gone up to over 420 ppm over the last century. Charles David Keeling, a scientist at the Mauna Loa Observatory on Hawaii, started recording CO_2 levels daily in 1958. The observatory's remote location and high altitude create a perfect site to take atmospheric measurements unaffected by human industrialization. And the graph of this information, know as the Keeling Curve,[14] shows us that the recent rise of CO_2 from 250 ppm to over 420 ppm lines up perfectly with the similar rise in global warming temperature.

Methane (CH_4) emissions have grown in recent years and was 11% of U.S. greenhouse gas emission in 2020. It comes from agricultural activities, waste management, energy use, and biomass burning. While 11% is a lot less than the 79% CO_2 emissions, we have to remember that different greenhouse gases absorb light differently and it turns out that methane is more efficient at trapping heat than CO_2. Scientists refer to this difference as Global Warming Potential (GWP).[15] Since CO_2 is the most common greenhouse gas, it's used as the reference gas and assigned a GWP of one. All other greenhouse gases are measured against it. For instance the 20-year GWP of methane is 85.[16] That means that a leak of one metric ton (tonne) of methane has the same climate impact as 85 tonnes of CO_2. So yes, we emit less methane, but it really, really hurts.

> *TRIVIA: We say "20-year" because different gases break down at different rates and their Global Warming Potential (GWP) changes over time. So scientists have to measure GWP for a fixed time period to get a fair comparison. The United Nation's IPCC likes to issue reports on 100-year GWP, but I find that number too remote. The 20-year number is what's going to effect us in our lifetimes.*

Nitrous oxide (N_2O) is the United States' third largest greenhouse gas emission at 7%, with a 20-year GWP of 264.(ouch!) Most of our nitrous oxide emissions come from agricultural activities, such as fertilizer use, but can also come from burning things and waste treatment.

The remaining greenhouse gas emissions are actually a group called flourinated gases, or F-gases. These are complex man-made gases used in industrial processes, refrigeration, and consumer products. They have an enormous impact on global warming. F-gases are only 3% of total U.S. emissions, but their individual GWPs can be in the tens of thousands. For instance, hydrofluorocarbons (HFCs), one of the most common gases used in air conditioners and refrigerators, have a 20-year GWP of 10,800. And sulfur hexafluoride (SF_6), commonly used in the electrical and semiconductor industries, has a 20-year GWP of 17,500.

Some of us may remember the ozone crisis during the 1970s-80s. Chlorofluorocarbons (CFCs)[17] from aerosol spray cans were found to be breaking down the ozone in our atmosphere, letting more ultraviolet (UVB) light reach the surface, increasing skin cancers, blindness and other related illnesses.[18] This was another case of complex man-made gases unwittingly changing the environment. If you can believe it, corporations voluntarily began removing CFCs from their products and their formal phase-out was codified by the 1987 Montreal Protocol,[19] the first universally ratified treaty in United Nations history and possibly the most successful environmental treaty ever. Since then, the atmosphere's ozone layer has shown a steady recovery and is on track to being back to normal by the middle of this century.[20]

So there, you have it. All the above shows us that increasing **Greenhouse Gases** are causing **Global Warming** which is creating **Climate Changes** that are triggering **Weather Disasters** which are destroying our lives, our infrastructure and our food systems.

We not only have to prepare for the warmer, more volatile environment we've already created, but we also have to stop increasing the damage to our climate systems. And that's where this Field Guide will help.

Once Again, With Politics

That was a lot of science stuff. We obviously have a very clear understanding of what's happening and how to fix it. So why don't we? Well, let's bring politics back into this for a moment. Fossil fuels are a massively rich and powerful industry. Multiple sources[21] report that oil & gas industry scientists predicted current global warming events with amazing accuracy as early as the 1950s; and that those industries chose to actively deny and negate their own research to maximize their long-term profits. As early as 1959, a Shell oil employee published an article in *The New Science*[22] tacitly acknowledging climate change by arguing that "Nature's carbon cycles are so vast that there seem few grounds for believing Man will upset the balance." In response to leaks of their own research, the oil & gas industry formed organizations like the Global Climate Coalition,[23] and the Information Council for the Environment,[24] that drew from the Tobacco Industry Playbook[25] – which protected tobacco revenues in the 1950s by denying the link between smoking and cancer – to successfully sow doubt about climate change and to reposition global warming as a theory, not a science-backed fact. This cult of Climate Denialism[26] successfully delayed climate action for decades and the gas & oil industry continues to spend $120-$180 million annually lobbying politicians and media[27] and another $84-$140 million in annual campaign contributions[28] to promote their climate-denial policies and ideas.

At this point, climate change and the global warming controversy[29] has become political rather than scientific. There's little doubt in the scientific measurements of growing atmospheric carbon dioxide taken daily at the Mauna Loa Observatory, or the increasing air temperature readings taken by weather stations around the world, or even the increasing number of billion-dollar weather disasters documented worldwide. The cult of denial has now focused on whether this undeniable change in climate is man-made (anthropogenic) and whether we can actually do anything about it. The deniers tell us that this is just a natural climate cycle, that there's really nothing we tiny humans can do about it, and that we might as well continue to enjoy our fossil-fueled comforts and profits for as long as we can.

Climate deniers like to point out that CO_2 is part of the natural cycle of carbon between plants and animals. When animals breath (respiration), they inhale oxygen (O_2) and exhale CO_2. When plants do their plant thing (photosynthesis), they absorb CO_2 and release O_2. These two interdependent systems of life – respiration and photosynthesis – are so perfectly balanced that their chemical formulas are mirror images. The formula for respiration is $C_6H_{12}O_2 + 6O_2 \rightarrow 6CO_2 + 6H_20$ and the formula for photosynthesis is $6CO_2 + 6H_20 \rightarrow C_6H_{12}O_2 + 6O_2$. Isn't that amazing? So its correct to point out that CO_2 is a natural part of the atmosphere needed for the survival of both plants and animals. And it's correct to point out that nature's carbon cycle has been thrown out of balance before by cataclysmic events like big asteroids crashing into the planet or extreme volcanic eruptions.

However, these reasons to deny man-made climate change don't explain the other greenhouse gases (methane, nitrous oxide, and man-made F-gases) that have hundreds to thousands of times more GWP than CO_2. And they don't address what happens when mankind deforests giant swaths of the planet at the same time as it invests heavily in animal agriculture. Isn't mankind increasing CO_2 from the billions of new livestock animals at the same time it's reducing the planet's ability to remove CO_2 through photosynthesis?

And what about cataclysmic carbon events? I don't think we've had any major asteroid hits or volcanic eruptions lately, but we did have an industrial revolution around 1800. The Industrial Revolution[30] started burning coal and petroleum for the first time, to drive new things like steam engines, electricity, and the internal-combustion engine. Does that raise any flags? Coincidentally, the line graphs of both atmospheric CO_2 and global warming begin their sudden upturn around 1800, in perfect unison with the industrial revolution.

So why would mankind burning things like coal and petroleum release more greenhouse gases than natural forest and prairie fires? First, consider the difference in magnitude. Mankind's ability to constantly burn things in millions of factories around the world far outpaces the frequency of natural forest or prairie fires. Second, consider what's being burned. Plants are made of cellulose which is 28% carbon by weight. Compare that to coal which is 75-85% carbon by weight. This higher carbon content is great for factories since coal burns hotter, but really bad for the atmosphere because it's releasing three times the carbon of a natural fire.

> *TRIVIA: Crude oil is 84-87% carbon by weight and natural gas is 65-80%. Hence, the argument that natural gas is better for the environment than coal. True, but less than 10% better? Really? Wind & Solar are pretty near zero-emissions once installed.[31] And that doesn't even touch on the enormous GWP of natural gas-related methane leaks.*

So the political football over climate change[32] continues to work its way up and down the field. This is really more about the liberals' belief in government oversight and social safety nets vs. the conservatives' belief in deregulation and market-driven solutions. Climate action takes a back seat to this ideological battle. Should we allow government to regulate or tax carbon emissions to fund climate recovery, or do we allow corporations to provide whatever products their customers want and let the market drive climate solutions?

There are certainly corporations that put the welfare of their employees and customers ahead of profits. But they are the exception. A corporation's legal responsibility is to maximize profits for its stock holders. If its goal was to lose money, it would be organized as a non-profit. We really shouldn't be surprised that oil & gas companies would engage in media campaigns and lobbying to maximize the sale of fossil fuels. That's what they do. The old adage, *caveat emptor* (buyer beware), applies to climate change as much as it does to used cars, and it's our responsibility to beware of green-washing and to make personal choices that assure the health, safety and welfare of ourselves, our communities, our children and future generations.

This Field Guide will help.

BY THE SECTORS

Climate is a big issue and it's easy to get lost in all the jargon and abbreviations. Where do you even begin to address something this complex? Even if we don't let it overwhelm us emotionally, it's far too easy to spend all our time and money digging out of the last disaster instead of preparing for the next. Governments spend so much time and money recovering from climate disasters that they never get around to fixing their root cause. They just keep putting out the fires without turning off the gas that's starting the blaze. The Institute for Public Policy Research (IPPR) calls this the "doom loop".[33] But, you know, it doesn't have to be that way.

There are clear, easy steps we can take to prepare for climate change; both the climate change that's already here and the climate changes coming down the road. The first step is to break this big, complex problem down into manageable sections; each with a group of small, clear actions that are easily taken and, together, help us secure our health and safety, save money, and position us for a safe and prosperous future.

The U.S. Environmental Protection Agency (EPA) has already done a great job breaking the climate crisis down into actionable sectors – what they call GHG Emissions By Economic Sector – and we'll use this as the map for our field guide going forward. These sectors can be looked at both nationally and globally; and we want to do both. The U.S. sectors are obviously important to those of us in the United States. And the global sectors are important because of all the stuff Americans buy from other countries.

Here's the EPA's 2020 breakdown of U.S. GHG Emissions by Sector.[34]

U.S. GREENHOUSE GAS Emissions
by Sector

27% – Transportation
25% – Electrical Power Production
24% – Industry
13% – Residential & Commercial (heating & cooking)
11% – Agriculture

In this case, bigger is badder. Bigger emissions mean that sector is causing greater damage to the climate and costing us more money. Let's compare this to the EPA's recent Global GHG Emissions by Sector,[35] which combines U.S. emissions with the rest of the world's.

Global GREENHOUSE GAS Emissions
by Sector

25% – Electrical Power Production
24% – Agriculture
21% – Industry
14% – Transportation
10% – Other Energy (extracting, processing & transport)
06% – Residential & Commercial (heating & cooking)

What does this tell us? It tells us where our actions, as Americans, can have the biggest impact. For instance, it's pretty obvious that U.S. transportation habits produce double the emissions of the rest of the world. The heating and cooking impacts of U.S. homes are also double the rest of the world. Agricultural emissions appear lower in the U.S. but keep in mind that a lot of the foods boosting emissions in the rest of the world are actually foods that are shipped to the U.S. for us to eat. So we can help with that.

This tells us that Americans get double coupons when we take climate action in the residential, transportation and agricultural sectors; or what I like to call Home, Travel and Food. Our emissions in these areas and the amount we pay for these unhealthy habits are so high that it's super easy for us to have a big impact with very simple changes.

So here's the sectors this Field Guide will walk us through.

POWER – The Electrical Power Production Sector
HOME – The Residential & Commercial Sector
TRAVEL – The Transportation Sector
FOOD – The Agricultural Sector
PRODUCTS – The Industrial Sector

PLUGGING IN

POWER – The Electrical Power Production Sector

Power drives pretty much everything in our lives. Power moves us around, protects us from the weather, feeds us, clothes us and lights us up at night. Power is fundamental to human life and good power choices can keep us healthy and secure. Bad choices don't.

Many of us are tapped into multiple power sources – gas pipes feeding natural gas into our home furnaces and stoves, coal-fired power plants generating electricity to light our lights, and the corner gas station pumping gasoline into our cars and yard toys. Most of us pay multiple utility bills, each with a fixed monthly administrative cost. Look at your utility bills and see how much of it is for fuel and how much is fixed costs you pay whether you use any fuel or not. Plus all those utility things like security deposits, taxes, meter fees, etc. that can add up when using multiple power sources. Wouldn't it be better if we could merge all our power options into a single account?

Obviously, you can't turn off your electricity. There are too many great things that only run on electricity. For instance, you can't plug your TV into a gas line. It's silly to even think about that, but is kind of amusing from a steampunk point of view. On the other hand, everything you have that runs on gas, CAN be replaced with something that you plug in. Today, there are electric options for everything, and converting everything to electric eliminates the administration costs and maintenance of that gas line. And the electric options are going to save you money in the long run, give you a much healthier environment, and help solve the climate crisis.

It really doesn't matter if you think climate change is man-made or not. Almost everyone agrees that greenhouse gases are causing global warming, and that global warming is creating climate changes that trigger weather disasters; weather disasters that destroy lives, infrastructure and food systems. All of that is pretty much hard science. And most of us agree that the biggest emitter of greenhouse gases that we have any control over is fossil fuels.[36] So, it doesn't matter which side of the climate argument you're on. All of us want to protect our families and finances from a very real weather crisis and the EASIEST way for us to do this is to stop burning things to power our lives; to replace fossil fuels with clean energy options.

Scientists are exploring lots of clean energy options, but for most of us the easiest options are wind and solar power. Utility companies can now build and maintain clusters of wind turbines and arrays of solar panels for less than it costs to operate their coal and gas-fired power plants.[37] Less! And most of them are doing it; rapidly replacing their aging coal-fired power plants with new wind and solar farms.

This is more profitable for the utility power companies because the fossil fuel technology peaked years ago, and new wind & solar tech continues to double its efficiency every few years. Plus, consider the ongoing operating expense. Coal and natural gas-fired power plants depend on a constant supply of fossil fuel extraction and transportation to keep running. Wind and solar, once installed, run on air movement and light from the sun; no extraction or transportation costs. That in itself should tell us where the future profits are. Can you imagine the glee of utility board members discussing the profits from installing wind or solar, then charging their customers for that free energy for decades to come?

Another thing is that we're running out of fossil fuels and what's left is harder to reach. We use to drill an oil well on our back 40 and get a gusher. Now we have to drill offshore or frack used-up fields to suck up leftovers. The growing costs of these fossil fuels are passed on to the power plants and eventually to us in our electric and gas bills. And it's only going to get more expensive. Stanford reports that ALL known oil fields will be gone by 2052; that natural gas goes away in 2060 and coal is gone by 2090.[38] Supply and demand, you know.

So let's look at the writing on the walls. They all say, "Fossil fuels are a dying power source." Fossil-fueled power costs more than wind and solar. Fossil fuels present multiple health and climate hazards. And we're going to run out of fossil fuels in the not too distant future. At its most basic, do you want to buy a new fossil-fueled home then replace its power systems before 2060? Is that a good investment? You may plan to sell your home before then but consider the impact of an outdated power system on your home's resale value. And what about a new car? General Motors has announced that it will discontinue ALL gas vehicles by 2035.[39] What's the resale value of your new gas car going to be in a few years?

Another thing to think about is your health. There's lots of research coming out of places like Harvard[40] and Stanford[41] that look at the toxic gases released in our homes by gas line leaks and burning fossil fuels in our furnaces and stoves. This has to be hurting us and our children. We've taken steps to remove lead paint from our homes. We need to take the same steps to remove fossil fuels from our homes.

In short, we want to be "plugged in", not setting things on fire. If it takes a spark or pilot light to get it going, replace it with something you plug in. Converting to all electric power is an investment in the resale value of your properties and a major investment in the health of your family. Plus it fights climate change. That's a trifecta of wins. Besides, it's a lot easier to plug something in than get that pilot light lit.

Many electric companies are ready now to sell you power from their new wind and solar farms. But even if you're still getting your electricity from an old coal-fired power plant, you should still convert. Shutting off the gas and going electric saves you money, increases the value of your home, and gives you a healthy environment. It's time to take the fossil out of fuel.

Bottom line: The following Field Guide chapters will show you how easy it is to electrify your home and car.

* * *

Talking Points

Regardless of whether climate change is man-made or not, eliminating fossil fuels is the easiest and cheapest way to reduce big weather disasters.

All fossil fuels will be used up in the foreseeable future, and what's left is becoming harder to mine and more expensive.

It costs less to build new wind and solar farms than to continue the operation of coal & gas power plants.

Burning fossil fuels in your home releases toxic gases that hurt you and your family.

Turning off the gas is one less utility bill to pay, saving you administrative & other fees, and gas line maintenance.

Action Items

Remove the Fossil from Fuel. Convert everything you own to

electric using the following Field Guide chapters.

Look for renewable energy options to power your home & office. Most utility companies now offer options to buy renewable energy (green tariffs).

If you own property get a rooftop or ground solar bid and produce your own electricity. Many regions now have net-metering and will PAY YOU for any extra power you produce.

SHELTERING IN PLACE

HOME – The Residential & Commercial Sector

When we talk about health and security, what could be more important than our home and office? We spend most of our time in these two spaces. We spend most of our income on these two spaces. We rely on them to protect us from the weather – both normal and extreme – and to protect our health from environmental hazards and toxins. Our home is where our children and loved ones live, both now and in the future. It is our shelter. Our castle. Our fortress.

Homes and offices are the fourth largest emitters of greenhouse gases in the U.S., and they have the biggest impact on our health and longevity of any built or natural environment. Fixing these living spaces is one of the easiest ways we can improve our health, save money and fight climate change.

This Field Guide is going to focus on your home, but many of these ideas apply to your commercial spaces. Changing things costs money, but the things we're going to talk about reduce your utility bills, so you get your money back over time while increasing the value of your property. You get to live in a healthier space while helping fight climate change. If you rent, you may not be able to do everything, but you can do a lot. And, if you're building a new home, the world's your oyster – you'll not only have lower utility bills, but you'll actually save money on construction.

Let's start with your existing home.

Your Lighting

What's easier than changing a light bulb? New bulbs use a fraction of the energy of older bulbs, last dozens of times longer, and present fewer health risks. You might find up to four different types of light bulbs in your home – incandescent, halogen, CFL and LED. Here's a quick overview.

Incandescent bulbs are the ones pioneered by Thomas Edison in the 1880s. Yes, they're that old. They make light by running electricity through a metal filament, usually tungsten, to heat it up to 4500°F and make it glow. Halogen bulbs,[42] often used as recessed flood lights, work just like the incandescents except that the tungsten filament is surrounded by quartz instead of glass, and the interior is flooded with a halogen gas which allows the tungsten to get even hotter while lasting longer.

Compact Fluorescent Light bulbs[43] (CFLs) are the ones with the spiral shaped frosted glass tube and are basically compact fluorescent tubes. Well named! They create light by running electricity through argon gas & mercury vapor to produce ultraviolet light that stimulates a fluorescent coating inside the tube to produce visible light. Light Emitting Diode bulbs[44] (LEDs) are basically a microchip the size of a speck of pepper that releases photons (light) when hit with electricity.

Incandescent & halogen bulbs waste 90% of your electricity producing unwanted heat. CFLs and LEDs don't rely on heating metal filaments so they don't waste that electricity. CFLs were "the bomb" in the 90s, but today's LEDs are better. Plus we've realized that having mercury vapor in breakable glass tubes scattered around our homes is really not a good idea.

> *WARNING: When you replace CFLs, be careful you don't break them and release the mercury vapor. And please take them to*

a responsible recycler. You don't want that mercury leaching from a trash dump into your water systems any more than you want it in your home.

So there are several reasons you want all your light bulbs to be LED. LEDs don't burn your fingers like incandescents and halogens. They don't give you mercury poisoning like CFLs. LEDs use a lot less electricity than any other bulb, lowering your utility bills, and they last over 20 times longer than traditional incandescent bulbs.

A nine watt LED bulb puts out the same amount of light as a 60 watt incandescent,[45] so you get the same light but save 86% on your electricity. The average life span of an incandescent is 1200 hours and a LED lasts 25,000. So one LED lasts longer than 20 incandescents. Even if you pay four times as much for that LED, it'll still be five times cheaper than 20+ incandescents.

Bottom line: Replace all your old bulbs with LEDs. Recycle the CFLs carefully. You can even get LED tubes to replace the old fluorescent lights in your office.

Your Heating & Cooling

Depending on where you live, heating and cooling can be one of your largest monthly expenses. And, with climate change, the growing weather extremes of hot and cold are making these expenses bigger. Changing to a modern heating-cooling system is going to pay for itself quickly, with savings only increasing over time. Plus, they're better for your health and fight climate change.

For most of history, heating relied on small open fires or stoves in each room. By the early 1900s, most homes had central heating. This burned coal, fuel oil or gas in a central boiler or furnace to create hot air, hot water or steam that was blown or pumped through ducts or pipes to vents or radiators in each room.

Nothing you really need to know.

The public saw "air conditioning" for the first time when the Missouri State Building was cooled for the 1904 World's Fair.[46] Air conditioning became commercially available in the 1930s, and was in most U.S. homes by the 1960s. Because air conditioners are all compressors and pumps, this was the first all electric way to change a building's temperature. It led to the invention of the electric heat pump which uses electricity to both heat and cool.

> *TRIVIA: Air conditioning works by compressing a refrigerant outside to squeeze out its heat, pumping that compressed refrigerant into a building, allowing it to expand and absorb inside heat, then pumping it back outside to be recompressed, dump its heat, and repeat. The electric heat pump works the same, but let's you choose whether you're compressing the refrigerant inside or outside of the building.*

Today, many homes have an air conditioner sitting outside, connected by flexible tubing to an indoor gas furnace. The furnace burns natural gas or liquid propane (LP) when heating is turned on, and acts as the air conditioner's heat exchanger (decompressor) when cooling is turned on. In both cases the hot or cold air is blown through metal ducts to vents in all the rooms. Some homes may have an all electric system that uses a heat pump outside (which looks like an air conditioner), connected by flexible tubing to an indoor heat exchanger (which looks just like a gas furnace) that blows hot or cold air through ducts to your rooms. There are lots of problems with both of these systems.

Let's consider your health. Burning natural gas or LP in your home releases toxic gases. Odorless ones. Building codes now require gas furnaces to be vented to the roof so the toxic gases can mess with you outside instead of inside. But the venting is not perfect and your gas furnace may be older than the vent code. You should check. There are also problems with small, undetectable gas line leaks. Your gas company adds chemicals to give odorless natural gas that "rotten egg" smell and make big leaks easy to sniff out. But small leaks still go undetected. A perfect set of gas pipes and furnace vents can protect you from these toxic gases, but is it worth the risk? And even if perfectly vented to the outside, these greenhouse gases still mess up our weather.

Another health hazard is the metal ducts that run from your furnace to all the vents everywhere in your home. Over the years, these accumulate dust, pollutants, and other allergens; not to mention the mice droppings, roaches, and occasional reptile escapee from your kid's terrarium. On the plus side, the mice eat the roaches and the snake will probably take care of the mice. But the allergens, bacteria and mold never go away. Yes, there are "duct cleaning" services, but the EPA says that, "Duct cleaning has never been shown to actually prevent health problems."[47]

A traditional heat pump eliminates the gas problems, but does it save you money? We've already compared the growing cost of fossil fuels to the shrinking cost of wind and solar. And we've talked about the higher resale value of an all electric home. But let's consider some other things. Your home may have hundreds of feet of ducts and the hot or cold air blown through these ducts is going to waste around 30% of your energy[48] heating or cooling the ducts before it even gets to the room vent. A central thermostat means you set the temperature in one room and the rest of the house has hot or cold spots. And you're spending money heating or cooling parts of the house you may not even use. Wouldn't it be great if you could just heat and cool rooms when they were needed?

So the ideal heating-cooling system would not burn gas, would not have any ducts, could set different temperatures in different rooms, and would use very little electricity to do all these things – that is, a ductless mini-split.[49]

Ductless mini-splits have been used in Europe and Japan for

over 50 years; and are used in over 90% of homes there today. Instead of connecting a heat pump to a central heat exchanger and blowing hot or cold air through ducts, the heat pump is connected directly to multiple small heat exchangers in different parts of the house (zones). Each zone circulates its own air, has a separate thermostat you can set to its own temperature, and offers other bells and whistles like programmable thermostats, remote controls and motion-detectors that adjust the temperature as you walk into a room. Because they're not wasting energy on ducts, ductless mini-splits can reach a Seasonal Energy Efficiency Ratio (SEER)[50] of 29.4[51] compared to a max SEER of 18[52] for ducted heat pumps. That's really good.

Ductless mini-splits are perfect for remodels as they're super easy to add to old buildings. You can just unplug your old furnace and cover up the duct vents. Or tear out the furnace and use that space for something fun.

Bottom line: Get rid of your gas furnace. Gas gets more expensive every year, is a big health risk and is bad for the climate. A ductless mini-split is going to pay for itself with lower utility bills, eliminate the health problems of burning gas and dirty ducts, and increase your comfort by letting you set different temperatures in different areas at different times – with a remote!

Your Kitchen

After heating & cooling, the kitchen is your next highest energy sink. Lots of appliances using lots of electricity. For the most part, just look for Energy Star labels[53] and select appliances that are more energy-efficient to lower your utility bill and help fight climate change. All Energy Star labels are earned by meeting minimum energy efficiency requirements set by the U.S. Environmental Protection Agency (EPA). If an appliance doesn't have an Energy Star label, then it doesn't meet these minimum standards. Look for an appliance that does.

Efficiency is super important in your large appliances like the dishwasher and refrigerator. And the way you use them counts. The Energy.gov website has an Energy Saver section that gives you all kinds of great tips on using kitchen appliances and other things.[54] Countertop appliances serve you and the environment best if they have multiple uses. Instead of say, a separate countertop microwave, convection oven and airfryer, why not invest in a single appliance that does all three? It'll be cheaper than buying three separate appliances, take up less space, and help with climate issues by reducing consumerism and manufacturing. But more on that in the Products chapter (What We Wear).

The several hundred-pound gorillas in your kitchen are your stove, range, built-in oven(s), cooktops, etc. Like many of your older systems, these may be burning gas. And, as you know by now, burning natural gas or LP releases toxic gases like carbon monoxide, not to mention the small gas pipe leaks that are venting things like methane and benzene into your room. Even when you're not cooking, small stove leaks and pilot lights continue to pump these toxins into your air. It's interesting that building codes are so worried about venting gas furnaces to the outside, but don't seem to care about the same toxic gases in your kitchen.

Back in the day, ranges were required to have a range hood vented to the outside. These days, most range hoods are recirculating, meaning they just pull the air through a screen – maybe a charcoal filter to remove some smell – then blow it right back into the room. Pretty sure that little metal screen is not removing toxic gases. And what about built-in ovens? No range hoods there.

A 2022 study published in Environmental Science & Technology[55] reported that 95% of the air samples taken near gas stoves contained at least 21 different hazardous air pollutants, including methane, benzene, toluene, ethylbenzene, xylene and hexane. Benzene in particular is known to cause anemia leading to leukemia and lymphoma cancer.[56] And a Stanford study[57] found that, "methane leaking from stoves inside U.S. homes has the same climate impact as about 500,000 gasoline-powered cars." Gas stoves are a problem.

Electric stoves, ranges and cooktops solve all of these gas-related health problems and help with climate issues. But electric ranges have their own problems. That big metal heating element in the oven – and the ones coiled up to make stovetop burners – work just like incandescent lights. They run electricity through a resistive metal filament so it will get hot and glow. Different metals give us a preference for glow in the bulb and hot in the stove, but they both still do the same thing. Turns out those toy Easy-Bake Ovens that cooked with incandescent light bulbs were right on point.

Like incandescent bulbs, electric stoves use way too much electricity for either our pocket books or climate issues. And they take forever to heat up and cool off, wasting even more power and making it really hard to do any fine, temperature-sensitive cooking. Fortunately, what LEDs did for lighting, Induction cooking does for our kitchens.

Induction stoves have been popular in Europe for decades. They use electromagnets to vibrate the iron or steel in your pan, which makes the pan get hot and cook your food.[58] Since it's the pan getting hot, not the burner, you buy a lot less electricity and your kitchen stays cooler. Electromagnets power up and down really fast, so induction can boil a liter of water in three minutes (40% faster than gas, 65% faster than electric) and chefs love the instant temperature control.

Another cool thing is the induction cooktop surface. Literally. It doesn't get hot. You can touch the burner at full heat and not get burned, and a cool surface means that spilled food doesn't burn. So cleanup is a lot easier.

> *WARNING: The induction burner stays cool, but the pan does not. Don't touch the pan; it's hot. And don't wear rings when cooking; induction will make your rings hot.*

Induction is available in ranges, cooktops, ovens, and portable units, which are perfect for renters. A portable induction burner costs about the same as a nice stainless steel skillet. And the prices of built-ins are comparable to similar gas and electric units.

Bottom line: Replace your gas or electric cooking with Induction, and make sure your other kitchen appliances have high Energy Star ratings.

Your Hot Water

Everyone loves a hot shower or a warm soak in the tub. But a gas water heater is just like a gas furnace, burning fossil fuels and releasing toxic gases in your home. Your gas water heater is legally required to be vented to the outside, just like your furnace. Most old gas water heaters are vented with a sheet metal duct sitting on top of the tank like a tin hat. What could go wrong?

Electric water heaters are pretty good these days. But, as with furnace ducts, a lot of the energy you buy is lost from hot water just sitting in the tank or radiating heat from those long hot water pipes running through your house. Why pay to heat water when you don't need it? And why stand in the shower running good water down the drain while waiting for the hot water to get there? You're paying to heat the water when you don't need it and you're paying for water that you throw away.

Tankless water heaters can reduce your utility bills 24%-34% over traditional tank heaters.[59] A tankless system instantly heats the water when it starts flowing through the pipe and shuts off when the water stops moving. You're not paying to keep a tank full of water hot when you don't need it.

There are lots of powerful tankless water heaters that you can plug right into your old tank water heater's pipes. But you're still wasting money heating those long hot-water pipes and standing in the shower waiting for the hot water to get there. Instead of putting a big tankless water heater in the basement, why not put several small ones near the faucets. They won't cost much more than one big tankless water heater and you won't have to heat the pipes or wait for hot water anymore. The only place you use hot water in the kitchen is at the sink – dish washers heat their own water – and you can plug in a small tankless water heater under the sink right next to the garbage disposal. A bathroom sink, shower and tub can share a small tankless water heater or, if you have bathrooms next to each other, a medium sized unit can provide hot water to all of them at the same time. This eliminates heating the pipes and you'll never have to wait for hot water again.

Bottom line: Replace your gas water heater with an electric one. Go tankless to save money. And, instead of a big tankless heater in the basement, put small ones near the faucets to save even more energy and never have to wait for hot water again.

Your Laundry

Modern washing machines offer a variety of energy-saving, low-water options. Save energy, save water, save money. Front-load washers are more efficient than top-load and they're gentler on your clothing. Look for a new washer with a good Energy Star rating and wash less often. Seriously. You'll save money and your clothes last longer when they're washed less.[60]

Replace your gas dryer with an electric dryer. It's healthier, good for your pocket book and for the climate. Look for Energy Star labels to select a good electric dryer. But even a good dryer uses a lot of electricity and pushes up your utility bills.

TIP: Keep track of your "laundry" days for a few weeks then go online and look at your daily electricity usage chart. You might be surprised by how big the spikes are on laundry days.

Whatever happened to clotheslines? The sun's UVB rays help sanitize your clothing and they have that wonderful, clean, "dried-outdoors" smell. You can't buy that in a bottle. If weather prevents or your rental doesn't have a clothesline, why not use a laundry drying rack? They're a common household appliance in Europe and Asia. You just unfold the rack and lay out your spin-dried clothes to dry inside. It works and helps humidify the house. Washers agitate and dryers tumble. Your clothes will last longer if they're neither agitated or tumbled. We have to wash, but a dryer? That's a choice.

Bottom line: Wash your clothes less and dry them on a clothesline or drying rack.

Your Fireplace

Do you have a fireplace? That's wonderful. There's nothing more relaxing than the crackling sound and sweet aroma of a hardwood fire. And nothing quite as bad as the hiss and smell of an old, dusty, asbestos-log gas fire. I know gas logs were a fad for a while, but there are some really decent electric fireplaces if you want a fake fire.

Bottom line: Replace those gas logs with a grate and enjoy a nice hickory or ash wood fire.

Why Weatherize

All these ideas to create a healthy, cost-effective, climate-savvy indoor environment don't help much if the outdoor weather is coming in through the windows and walls. New windows can be expensive but pay for themselves over time. New storm windows are a cheaper way to keep the outdoors out and the indoors in. And sealing air leaks with some old-fashioned caulking and weather stripping is always a great way to get the most from all the ideas in this Field Guide.

It Takes Two

Two what? To do what? If we want to protect ourselves from even more disastrous climate disasters, we have to do two things. We have to reduce our greenhouse gas emissions and we have to find a way to remove existing greenhouse gases from the air – a process called sequestration.

All the ideas in this Field Guide help you save money, position yourself to deal with bigger temperature extremes, and cut back your household's greenhouse gas emissions. But the climate has already started changing. Glaciers and ice caps are melting. Forests and scrublands dying. And these changes add their own greenhouse gases to the problem. Even if we could magically, instantly stop all of our human greenhouse gas emissions, the environment has already tilted enough that climate change is going to continue picking up speed all by itself. To stop this, we have to "sequester" existing greenhouse gases from the air.

Fortunately, there's a very simple way you and I can sequester. Remember the section in the intro that explained respiration and photosynthesis? Oh, you skipped the intro? Well, there was this great section that explained how animal respiration adds greenhouse gases to the air and how plant photosynthesis removes them and how, in a healthy world, they balance each other. The bottom line is that plants "sequester"; they remove greenhouse gases from the air. We need more of that to stop climate change from getting worse.

TIP: It takes two. We have to both reduce greenhouse gas emissions and sequester existing greenhouse gases too.

So what about outside our homes and businesses? That's where most plants live and that's where we can increase sequestration by planting more plants! Stopping the deforestation of the Amazon rainforest and stopping commercial logging in U.S. National forests are important things, but so is turning our yards into gardens. Into sequestration engines. Into home forests. We already spend lots of time and money on our yards. What if we could save money, save time, and sequester too?

Let's start by taking the fossils out of fueling our yard toys. Do you have gas-powered lawn mowers? Edgers, blowers, trimmers? There are better electric models for all of these. But let's take it a step further. Do you really need these tools? How much time and money do you spend mowing and trimming? How much do you spend on lawn fertilizers and weed killers?

Grass sequestration is good, but "lawns" create more greenhouse gases than they sequester, which is not good. If you replace your lawn with an unmowed ground cover, you save money, save time, you won't have to maintain a mower or pay for yard service, and your yard will be removing greenhouse gases instead of making them. That's a lot of wins! And there are a lot of different styles of ground covers you can use to replace your lawn; and lots of new city codes that let you do it.

Ground covers like vinca minor (periwinkle) are hardy, grow 3-6" tall, squeeze out weeds and are scattered with lovely purple flowers in the spring. Natural prairie grasses make ground covers needing little or no maintenance. I've seen suburban yards fully covered with local wildflowers and even urban yards that have been allowed to revert entirely to scrub. All of these are great ways to sequester, and give you a variety of yard styles to choose from

– from urban wilderness, to fields of flowers, to waving prairie grasses, to prim periwinkles.

With all the time and money you save ditching the lawn, you can move to the next level of home sequestration – filling your yard with bushes and trees. There are lots of options. Clumps of bamboo are beautifully decorative and grow amazingly fast. Fruit and nut trees can reduce your food bill without the labor of a traditional garden. It's always great to support local native plants but, with the climate changing, native today may not be native tomorrow. Embrace the change and plant for the future.

WARNING: Depending on your area, things like vinca and bamboo can be so hardy they become invasive species; and may be best suited to urban yards where they can't spread to the wild or farm fields. Check with your local State Department of Conservation.

The ultimate home sequestration engine is what I call the "home forest." This is a deliberate choice to plant a highly dense, multi-layer strip of vegetation that is a super-sequestrator. Research in India has demonstrated that over 300 trees can grow in the area of six standard car parking spaces.[61] These trees are planted very close together with a combination of short, medium and tall varieties that create a dense wall of foliage which provides privacy and makes a great sound barrier.

TIP: A home forest is more than just trees. It is undergrowth; the bushes, shrubs and young trees that live under the canopy.

With the price of food these day, a garden can more than pay for itself. Yes, a traditional vegetable garden takes some labor, but many people find that gardening helps with stress and depression; not to mention that a little exercise is always good. Plus, fresh garden produce and the fruit and nuts from your home forest are going to be a lot tastier than the grocery store stuff.

Bottom line: Replace all gas yard tools with electric ones. Get rid of mowed lawns and landscape with groundcover, bushes and trees.

Renters

If you rent, the landlord probably won't let you make big changes. But there are still ways to create a healthier environment that is climate-savvy and saves you money on your utility bills.

It doesn't matter if you own or rent, new LED light bulbs are safer, healthier and save money. You can always put the old bulbs back and take the LEDs with you when you move.

You don't have to use your rental's old gas or electric stove. Just shut the stove off, put a small piece of plywood on it (to avoid scratches) and set a portable induction burner on that. And you can get a super-efficient portable convection/air fryer/microwave to replace that old oven. These countertop appliances will pay for themselves by reducing your utility bills, and you can take them with you when you move.

Rentals can also be weatherized to reduce your utility bills. The U.S. Department of Energy manages nationwide Weatherization Assistance Programs[62] (WAP) that can help you with instructions and supplies. But even something as simple and do-it-yourself as duct tape can stop air blowing through walls and windows. Taping aluminum foil on glass can reflect the sun away when it's hot outside, and be pealed off to let that warm sunlight in when it's cold.

Bottom line: Renters can save by owning their own LED lightbulbs and countertop cooking appliances.

New Construction

If you're building a new home, the things we've been talking about will not only lower your future utility bills; they will cut your constructions costs! If you're building to sell, these things cut cost and give you a long list of climate-savvy talking points to share with your buyers.

A ductless mini-split save lots of money because you don't have to pay for furnace duct materials, installation or vent covers. You don't even have to build a furnace room.

Kitchen venting is not required for induction cooking, so you don't need to install range hoods or other vents. Induction releases zero toxins and doesn't heat the air like gas or electric.

Install small tankless water heaters near the faucets and you only need to run one cold water pipe throughout the house, instead of two (hot & cold). This cuts your water pipe materials and labor in half. Your single cold water pipe branches at the end to feed both the tankless heater and the cold water tap. Even in a large bath with a shared tankless heater, you're only going to need 10-20 feet of hot water pipe instead of hundreds to run hot water from the basement.

Fireplaces are expensive and waste energy, so save your money. A nice firepit in the backyard is much cheaper and more enjoyable.

And while we're talking about yards, skip the sod and seed with a nice ground cover. You're creating a lower maintenance home that's better for the environment. Take the money you save and invest in native bushes and trees. Strategic placement provides shade and wind breaks to ease the load on your ductless mini-split and to help block weather extremes. Space large trees far enough away from the building to avoid branches falling on the roof. Extra home value if you plant some fruit and nut trees. Even with climate change, outdoor spaces like balconies and decks add value to your home. And don't forget to set aside spaces for possible gardening and composting.

New construction is the perfect time to weatherize. That extra effort to insulate and create an airtight space lowers utility bills forever. And don't get cheap with your windows. The U.S. Department of Energy reports that 30% of heating-cooling energy is lost through bad windows.[63] Good windows let the sun in to create a healthy space and provide passive solar heat that lowers your heating bills. Good windows keep outdoor temperatures out by being airtight and putting multiple layers of glass between you and the weather. Top-end windows let the sun shine through while blocking 97% of that outdoor hot and cold. And not all windows have to open. Stationary glass is cheaper and more airtight.

> *WARNING: One window per room should open to get you out in emergencies and let that occasional perfect weather in. Check your local building codes.*

Bottom line: New construction is the BEST time to build all these great features into your home. You get all the health and cost benefits we've described and save bucket-loads of money on construction. Plus, they make a great set of talking points when you're ready to sell.

* * *

Talking Points

CFL bulbs contain mercury vapor that is harmful if the bulb is broken.

Incandescent and halogen bulbs heat their filaments to over 4500°F and can cause serious burns.

LED bulbs use 90% less energy than incandescent and halogen bulbs.

LEDs last 25 times longer than incandescents and halogens so you buy fewer bulbs.

Gas furnaces release toxic gases through improper venting and small gas line leaks.

Furnace ducts lose 30% of your heating-cooling before it ever reaches the room.

Furnace ducts accumulate dust, allergens, pollutants, insects, rodents and are a breeding ground for bacteria and mold.

Central heating-cooling systems throw money away by heating-cooling unused rooms and forcing one temperature everywhere in the house.

Ductless mini-splits save money by using less energy than ducted heat pumps or gas furnaces.

Ductless mini-splits eliminate the need for ducts, making construction cheaper and the home healthier.

Ninety-five percent of air samples from gas stoves contain 21+ hazardous toxins.

Energy Star labels show you how much electricity a kitchen appliance will save.

Multi-use appliances are more efficient and cheaper than buying multiple single-use appliances.

Induction cooking uses magnets to heat the skillet instead of the air.

Induction cook surfaces stay cool when cooking.

An induction burner can boil a liter of water in three minutes.

Tankless water heaters use 30% less energy than hot water tanks.

If you have small tankless water heaters near the faucets or shower, you'll never wait for hot water.

If you put small tankless water heaters at the faucets in your new construction, you only have to run one cold water pipe throughout the house instead of two (hot & cold).

Whatever happened to clotheslines?

Drying racks dry you clothes inside while humidifying your house.

When was the last time you used your fireplace?

It takes two to stop climate change. We have to stop greenhouse gas emissions and we have to remove existing greenhouse gases from the air.

Photosynthesis removes greenhouse gases from the air. More plants means less greenhouse gases, means fewer climate issues.

Action Items

Replace all your lightbulbs with LED bulbs.

Replace your old heating-cooling system with a ductless mini-split.

Switch to induction ranges, cooktops and ovens.

Replace your water heater with small tankless heaters installed near the faucets.

Dry your clothes on a drying rack or outside on a clothesline.

Toss the gas logs and enjoy a real hickory or ash wood fire.

Replace your mowed lawn with groundcover.

Plant a home forest.

HOW WE GET THERE

TRAVEL – The Transportation Sector

Mobility is our ability to get from one place to another. This could be anything from visiting a friend, a neighbor or relative, to picking up groceries at the local market, to commuting to work, to visiting the Grand Canyon, to attending the United Nation's next Intergovernmental Panel on Climate Change (IPCC). Getting there is not only the largest emitter of greenhouse gases in the U.S., but probably your personal biggest expense after your home – buying that car or truck, paying property taxes, insurance, maintenance, and fuel. And how you "get there" has a direct impact on your health and longevity, not to mention that of your family, friends and co-workers.

Consider this list of mobility options ranked from lowest greenhouse gas emissions to highest.

CLIMATE Impacts of Mobility Options
ranked from Lowest GHG Emissions to Highest

Walking & Biking
Electric Buses & Lightrail
Electric Vehicles
Gas Buses
Gas Vehicles

Walking has the lowest greenhouse gas emission of any mobility option; zero if you don't breath too hard. Most cities have an extensive network of sidewalks and pedestrian right-of-ways.

And walking through nature is its own pleasure. The Federal Highway Agency reports that 28% of ALL auto trips taken in the U.S. are less than one mile.[64] Most people can walk a mile. That's around 2000 steps, a lot less than the doctor-recommended 10,000 steps a day you're trying to rack up on your step counter.

Biking – the pedal kind – has the next lowest greenhouse gas emission, taking into account the bicycle's manufacture and delivery. And it's a great source of cardio, vitamin D, etc. There's a growing network of urban bike lanes, not to mention the popularity of mountain bikes for off-road cycling. The Federal Highway Agency also reports that 50% of ALL auto trips taken in the U.S. are less than THREE miles. Anyone can bike three miles. The average speed of a leisurely bike ride is 14 mph,[65] so you're talking maybe 12 minutes for that three mile trip – probably less time than it takes you to get in a car, park, etc.

The most climate friendly form of passive transportation (the kind where you sit there and let a motor do the work) is electric mass transit, which can be electric buses or lightrail systems like streetcars, subways, and elevated trains. Electric mass transit combines the climate-friendly electric motor with the efficiency of a shared ride.

Electric vehicles (EVs) are the best climate option when walking or biking is out and when electric mass transit doesn't go where you need to be. A medium EV emits around 53 grams of greenhouse gas per kilometer when charged from a fossil-fueled power plant, and practically zero if charged with electricity from a wind or solar farm.

Gas buses comes in a lot of different flavors and are readily available. Riding a diesel bus reduces your personal greenhouse gas emissions 45% over driving a medium gas car.[66] Multiply this by the number of cars mass transit can take off the road and you can see that this is a great alternative to that old gas burner parked outside.

TRIVIA: An average bus can replace 40 single-passenger cars. A small lightrail can replace 150 single-passenger cars.

Now let's consider our list of mobility options ranked from healthiest to least healthy.

HEALTH Impacts of Mobility Options
ranked from Healthiest to Least Healthy

Walking & Biking
Electric Buses & Lightrail
Electric Vehicles
Gas Buses
Gas Vehicles

Active transportation (walking & biking) is understandably the healthiest way to go. These are the types of activities constantly recommended by doctors. They're proven to reduce heart disease, diabetes, and depression. We all should walk and bike more.

Electric mass transit (buses & lightrail) provides a quiet, peaceful way to get to work or across town. The time spent relaxing, reading a book or watching the world go by, lowers stress and helps fight depression. Electric vehicles aren't quite as relaxing, because you're probably driving and have to, well, drive. But they're still quiet and free of the exhaust fumes and carbon monoxide that comes with the gas ride. Gas buses at least provide a more restful ride than driving, although the noise and fumes make any electric mass transit option seem better.

Finally, let's rank our list of mobility options from most savings to most expensive.

COST Impacts of Mobility Options
ranked from Least Expensive to Most Expensive

Walking & Biking
Electric Buses & Lightrail
Electric Vehicles
Gas Buses
Gas Vehicles

Are you starting to see a pattern?

Walking is the cheapest form of mobility. You just stand up and go. Biking requires the purchase of a bicycle, but this is usually a one-time expense with minimal maintenance cost. You can travel thousands of miles with one small investment in a good bike. Active transportation also saves you taxes since pedestrian and bike lanes are cheaper to build and easier to maintain than car lanes.

Riding mass transit, whether it's electric or gas, can save you quite a bit of money compared to the fuel and upkeep of a personal vehicle. And many cities now provide free mass transit, making your cost comparison pretty simple. New electric buses can go over 1000 miles on a single charge.[67] And cities are discovering that upgrading their aging bus fleets to electric actually lowers their operating expenses. What's not to like?

Electric vehicles (EVs) are the most cost-effective personal car or truck you can own. Retail prices (MSRP) for electric vehicles are the same as old-fashioned gas cars and, with tax rebates, can be even cheaper. But where electric vehicles really shine is saving you money on maintenance. Electric vehicles don't have mufflers, starters, oil pumps, spark plugs, gaskets, or a lot of other stuff that needs constant maintenance in a gas vehicle. So none of that is ever going to need to be replaced or repaired in an EV.

Modern electric vehicles can go hundreds of miles on a single charge. Remember that 50% of all trips in the U.S. are less than three miles? How far is your commute? Most of us will be able to commute for several days on a single charge. And, you'll probably be plugging in your car each night anyway.

> *TRIVIA: There are three levels of EV chargers.[68] Level 1 is a normal house outlet. It charges pretty slow, but if your car is going to be garaged 12 hours a day, it gets the job done. Level 2 charging is basically what you plug your 240v electric range, dryer or hottub into. Charges around five times faster. Then there's level 3, which uses direct current instead of U.S. alternating current. These are what you see at all the public EV charging stations. They charge 40-80 times faster than level 1. You can find everything you want to know about charging EVs at the U.S. Department of Energy's Electric Charging At Home site.[69]*

Electric vehicles now out-perform gas vehicles on every front. A U.S. Department of Energy study found that scheduled maintenance on an average EV costs 6.1 cents per mile compared to a cost of 10.1 cents for an average gas vehicle.[70] A 2018 study at Michigan's Transportation Research Institute found that the average cost to operate an EV in the U.S. was $485 per year, compared to an average cost of $1,117 to operate a gas vehicle.[71] The Union of Concerned Scientists found that, over the lifetime of your car, the average emissions of an EV is 28 metric tons compared to a gas vehicle's 57 metric tons.[72] The Department of Energy reports that EVs convert 77% of electricity's energy to forward motion (or reverse), while gas vehicles only convert 12%-30% of gases energy.[73] At the time of this writing, the Department of Energy's FuelEconomy.gov site shows that a 2023 Chevy Bolt EV has a base MSRP of $25,600.00, gets the equivalent of 120 MPG and has a range of 259 miles on a single charge.[74] Again, what's not to like?

Keep in mind that we're talking about electric vehicles, not hybrids. Hybrids are super complex with both gas and electric engines onboard. They were great when they came out because they introduced us to the potential of electric vehicles, they let us use normal gas stations, and they raised our MPGs a little. But, with the longer distances new electric vehicles can go, EV charging stations popping up everywhere, and an electric vehicle's much lower maintenance cost, what's the resale value of a new hybrid going to be in a few years? Besides, isn't a hybrid kind of like eating a fast-food burger and fries; then ordering a diet soda? Think about it.

Bottom line: Walk and bike more. Your heart and pocketbook will thank you. For longer trips, grab a good eBook and hop on the bus or lightrail. When you need to own, electric vehicles are the way to go.

Long Haul

If 50% of all trips are less than three miles, I imagine 99% of all trips are in-town. Walking, biking, EVs, local buses and lightrail are great for in-town. But what about the long haul? Those domestic business trips or vacations? What about international travel?

You generally have three options when traveling out of town. Drive, fly, or hop on a train. Flying has the most emissions, but not by much. Flying 500 miles emits 201 kg of greenhouse gas and driving a gas car that same distance emits 180 kg.[75] A little better but not a lot. If you drive an EV that 500 miles you emit 59 kg. Definitely better. But if you were to hop on a train, you'll only emit 27 kg of greenhouse gas. Wow!

We don't hop on trains much in the United States. They're not that reliable. Other places like China, Japan, and Europe have reliable, modern, high-speed rail; most of it electric. There, a 500 mile trip (which they'd call 805 km) on something like the EuroStar Train would only emit 4.8 kg of greenhouse gas.[76] That's the next best thing to having Scotty "beam you up."

> *TIP: If you're traveling in the E.U. grab a multi-day EuroStar pass. See what modern transportation is really like.*

So your best out-of-town mobility option is a nice train ride. Sit back and enjoy the view. Your next choice is a nice road trip in an electric vehicle. After that, you're down to drive gas or fly. Which is worse? A plane produces most of its greenhouse gas during take-off and landing, so long distance flights emit less per mile than short flights. If your trip is less than 620 miles, drive the car. If it's longer, then fly. And, if you're taking a family trip, the car loaded with people is always going to be cheaper and have less emissions than flying.

A few years from now? Maybe not so much. Regional flights are already moving to electric planes. Hawaiian Airlines is investing in all-electric, 100-passenger planes to fly 130 daily inter-island trips by 2028.[77] United Airlines has ordered hundreds of electric planes from Archer Aviation to provide air taxi services between their airport hubs and downtown areas starting in 2025,[78] and has ordered 100 larger electric planes from Heart Aerospace to start flying regional routes by 2030.[79]

Research at Airbus and other aircraft manufacturers is already putting larger hybrid planes in the air.[80] Jet engines to take-off and land. Electric engines to taxi and maintain level flight. Just as the hybrid cars launched the electric car movement, hybrid planes are going to be the pathway to long-distance electric flight. A couple decades from now, we'll all be flying electric across the world.

Bottom line: Drive your electric car on trips in the U.S. and use the electric rail systems in other countries.

* * *

Talking Points

Walking and biking are super cheap, super healthy and great for the environment.

The U.S. Federal Highway Agency say 50% of all trips in the U.S. are less than three miles. And 28% or all trips are less than one mile.

Electric vehicles don't have mufflers, starters, oil pumps, spark plugs, gaskets, etc., so maintenance is cheap and easy.

A full bus takes 40 single-passenger cars off the road. A full lightrail can take 150 single-passenger cars off the road.

Action Items

Explore your neighborhood's pedestrian and bike lanes.

Just for the fun of it, check out a bus or lightrail schedule for your work commute. Who knows ...

Test drive an electric vehicle. Not a hybrid, an EV.

Next trip, take the train.

WE GOTTA EAT

FOOD – The Agricultural Sector

Yes, we do. Food is not only one of our most basic needs, it can be one of our most sublime pleasures; filling our senses with subtle savory flavors, aromatics and spice. Unfortunately, the skyrocketing price of food forces many of us to eat the cheapest, least nutritious, far-from-sublime options available. We're forced to eat a sad diet that hurts our health, hurts the environment and, to add insult to injury, hurts our pocketbooks.

Food has become the second leading cause of climate change after power production. And food is the leading cause of our most common ailments, like diabetes, heart disease and cancer. The cost of shipping food thousands of miles raises our food bills and the resulting short shelf-life ends up with us tossing out 40% of our rotting groceries, costing us even more.

Industrial farming has ended up removing much of the nutrition from our food. We spray herbicides & pesticides to sterilize the fields, then we pour fertilizer on the field to get a plant to grow. We repeat this process over and over, and the soil dies. All of the great organic compounds and micro-nutrients are gone. Our health suffers and food tastes horrible.

Commercial Animal Farming Operations (CAFOs) are just as bad. Mass produced animals that are raised in tiny spaces, chain-fed antibiotics to slow mutating diseases, then live-shipped around the world to domestic slaughter houses. There's nothing nutritious or healthy about that. It's given us the pandemics of

1918 and 2019; viral outbreaks like MERS, SARS, and Avian Flu; and is now producing antibiotic-resistant superbugs.[81]

The real question is how long can the planet sustain these industrial food systems. With the world's growing population facing climate change, drought and famine,[82] the international transport of food is a serious geopolitical problem. China invaded Tibet to secure the Himalayan watershed[83] for drinking and agricultural use, and is aggressively pushing into international and foreign territorial fishing waters[84] to expand their food supplies. Russia's invasion of Ukraine has highlighted the world's dependence on foreign grains.[85] International conflict in Northern Africa is a struggle over the dwindling water resources of the Nile river.[86] And similar conflicts are growing among U.S. states over the Colorado river.[87]

Wow! That's a lot to unpack. But there are solutions to all of this. Simple dietary choices can lower our dependance on foreign food chains and put healthier food on our plates. And a better understanding of dietary styles will give us the tools to take back control of our lives, our health, our pocketbooks and our environment. Let's look at the different ways the food on our plate matters.

Food & Climate Change

The EPA reports that agriculture is 24% of global greenhouse gas emissions and the U.S. has to shoulder a lot of that. We're the ones paying to import trillions of tons of off-season food. The USDA reports that the U.S. imported almost $23 billion worth of fruit and $16 billion worth of vegetables in 2021.[88] That's a lot of foreign farming, packaging, and rush long-distance shipping to stock local grocers with off-season produce. And we imported over $13 billion worth of meats. Makes you wonder. Would Brazil be burning the Amazon forest to make more pasture if we weren't buying so much meat?

There are lot of studies looking at food and climate change. And it turns out that what we put on our plate matters. A lot! The Drawdown Project created a ranked list[89] of the "most substantive solutions"[90] to drawing down greenhouse gases. Twenty percent of the solutions are food-related. Look at just the top 24 solutions, and 46% are food-related. Are you getting the idea that food choices are ... well, substantive? The Drawdown Project found that the #1 solution to climate change is "Reduced Food Waste" and the #2 solution is "Plant-Rich Diets".

Another study published in the Journal of the Proceedings of the National Academy of Sciences (PNAS) looked at how much we'd reduce greenhouse gas emissions if we changed our dietary style.[91] They found that everyone changing to a doctor-recommended "healthy global diet" (HGD) would reduce greenhouse gas emissions over three billion metric tonnes ($gtCO_2e$) a year. Changing to a vegetarian diet would reduce emissions over seven billion metric tonnes and changing to a vegan diet would reduce emissions eight billion metric tonnes a year.

GigaTonnes (gt) of Greenhouse Gas REDUCED
per Year by Dietary Style

----

3 gt – HGD
7 gt – Vegetarian
8 gt – Vegan

A study from the Institute of the Environment at UC Davis looked at the greenhouse gas emissions of various foods. They found that a typical serving of beef produces 330 grams (g) of greenhouse gases, a serving of chicken produces 52 g, fish produces 40 g, vegetables 14 g, and lentils 2 g.[92]

Grams (g) of Greenhouse Gas PRODUCED per Serving

----

330 g – Beef
52 g – Chicken
40 g – Fish
14 g – Vegetables
2 g – Lentils

A comprehensive study by the University of Oxford, looked at 40 food products from 38,700 farms in 119 countries – basically 90% of everything eaten.[93] They found that beef produces up to 105 kilograms (kg) of greenhouse gases per 100 grams (g) of protein; in comparison to something like tofu, which produces less than four kg. Different farming practices produced different emission levels, but the lowest emissions from beef were still higher than the highest emissions from pork, chicken or fish. Confusing? Let's summarize Oxford's findings this way. To grow 100 grams of protein, beef produces up to 105 kg greenhouse gas, pork produces 14 kg, poultry 12 kg, fish 12 kg, tofu < 4 kg, beans < 2 kg, and peas < 1 kg.

Kilograms (kg) of Greenhouse Gas PRODUCED per 100 Grams of Protein

----

105 kg – Beef
14 kg – Pork
12 kg – Poultry
12 kg – Fish
< 4 kg – Tofu
< 2 kg – Beans
< 1 kg – Peas

Is this a good place to say that Americans really love their beef? A plant-rich, healthy global diet (HGD) doesn't mean you have to be vegan. You don't even have to be vegetarian. A plant-rich HGD means you eat less meat with more of those doctor-recommended servings of fruits & vegetables. But more on that in the Health & Longevity section.

> *TRIVIA: A doctor-recommend healthy global diet (HGD) includes 4-6 more servings of fruit and veggies daily, cuts your six ounce serving of meat down to four, and cuts your food-related greenhouse gas emissions by 50%.*

Livestock accounts for 58% of agricultural greenhouse gas emissions. That's a lot. All of the studies agree that the greenhouse gas emissions of growing livestock are a magnitude higher than growing plants, and that the emissions of growing beef are a magnitude higher than any other meat. If you're concerned about climate change, reducing the meat in your diet - particularly the red meat - will lower your greenhouse gas emissions.

> *TRIVIA: The typical meat-heavy western diet produces 10 times the greenhouse gases of a vegan diet.[94]*

Bottom line: Beef produces more greenhouse gas emissions than other red meats like mutton & lamb; which in turn produce more greenhouse emissions than white meats like poultry, pork and fish.

Food And Land Use

Remember the Sheltering In Place chapter when we were talking about sequestration? That it takes two things to stop climate change and that one of those is growing more plants to scrub greenhouse gases from the air? That's why land use is important. Every time we chop down a forest or burn off a patch of scrub, we're reducing the planet's ability to remove greenhouse gases and increasing climate change.

We need farming, but it uses a lot of land which could be covered with forest and scrub that was fighting climate change. How do we provide more calories to a growing world population but use less land? We need to think about how our crops are being used. There are food crops and there are feed crops. If people eat it, it's a "food" crop. If we feed it to livestock, it's a "feed" crop.

It takes 4.45 acres of land to feed the livestock needed to feed one person on a meat diet for one year. It takes one-third acre of land to feed a person on a vegan diet for that same year.[95] That's almost fourteen times as much land for a person eating meat. What do you think would happen if we used more of our crop land for food instead of feed? You can grow 37,200 pounds of corn on one acre,[96] or you can raise 354 pounds of beef[97] on that acre. How many more people could we feed if we ate the corn instead of the cow?

Around 83% of the world's farmland is used to raise livestock and their feed crops. But livestock only provides 18% of the world's calories.[98] The other 82% of our food calories come from food crops growing on the remaining 17% of world farmlands. Think about that. Less than 20% of the world's farmland is providing over 80% of the our food calories. Acre for acre, food crops provide a lot more calories than livestock.

What if we doubled the amount of land we use to produce food crops? Say, use 33% of our farmland for food crops instead of 17%? That would almost double the calories from food crops to 159% of current numbers. This would leave 67% of all farmlands to raise livestock and lower livestock calories from 18% to 15%. We're still using two-thirds of all farmland to produce meats, but we've increased available food calories to 174% of current numbers. That would go a long way toward fighting hunger. That would go a long way toward addressing world population growth.

Here's another scenario. What if we went 50/50? Used 50%

of all farmlands for food crops and the other 50% for livestock? That would increase calories from food crops to 241% of current numbers and lower calories from livestock to 11%. A 50/50 farming split would increase available food calories to 252% of current numbers!

> *TRIVIA: Twenty-six percent of the Planet's ice-free land is used for livestock grazing and 33% of croplands are used for livestock feed production.[99] So 59% of all available land is used to raise meat.*

Since we're talking about climate change, let's consider a third scenario. Let's use 33% of our farmlands to produce food crops, leave 33% of our farmlands raising livestock, and repurpose the remaining 34% of farmlands into forests and parks. This scenario increases available food calories to 166% of current numbers, and puts over four billion acres[100] into service scrubbing greenhouse gases from the air.

Bottom line: We use most of our farmland to feed livestock. If we reduce our meat consumption to a doctor-recommended healthy global diet, we'd have a lot more land available to feed people.

Food And Water Use

Globally, agriculture uses 70% of all available fresh water.[101] Different foods obviously need different amounts of water. This list shows the gallons of water it takes to produce one pound of each food.[102]

Gallons of Water Required
per Pound Food

----

22 – Tomato
36 – Broccoli

36 – Potato
40 – Strawberries
44 – Eggplant
68 – Oranges
76 – Lemons
128 – Milk (cow)
380 – Cheese
392 – Eggs
520 – Chicken
668 – Butter
720 – Pork
1852 – Beef

Once again, we see that beef stands out, even among the meats. And we need to point out that we buy cow milk by the gallon, not by the pound. A gallon of milk weighs around 8.6 pounds, so do the math ... it takes 1100 gallons of water to make ONE GALLON of milk.

> *TRIVIA: It takes 660 gallons of water to produce one quarter-pound hamburger.[103]*

Food And World Population

World human population is now over eight billion. The rate of growth peaked in the 1960s and has been slowing since, but the human population is still growing.[104] We need to produce more food calories to feed everyone. And with climate change destroying our croplands, we need to produce even more food calories per acre.

The world livestock population is growing faster than humans. It's now over 21 billion. For instance, the world mammal population is now 60% livestock, 36% human and just 4% wild. The world bird population is now 70% chicken and other poultry, and just 30% wild.[105] This loss of biodiversity is itself an ecological disaster, but we also have to think about how much food and water it takes to sustain all this livestock.

We've already talked about livestock using more land and water than food crops. What about the calories and protein? Which will feed more people? A feedlot uses six pounds of corn to produce one pound of beef. What's the equivalent nutritional value of these two foods? Remember, it takes six pounds of corn to grow one pound of beef.

Nutritional Value of Beef
per Pound

1152 – Calories
91 g – Fat
0 g – Carbohydrates
78 g – Protein

Nutritional Value of Corn
per Six Pounds

2340 – Calories
32 g – Fat
518 g – Carbohydrates
88 g – Protein

We'd get double the calories and 13% more protein if we'd eat the corn instead of the cow.

Another way to look at food is how much energy it takes to grow. Something may provide lots of calories and protein but if it uses too much energy, we're going to starve. The best foods are the ones that take less energy to produce more calories.

Corn provides 390 calories per pound and takes 0.43 kilowatt hours (kWh) of energy to produce that pound.[106] If we divide corn's calories by its kWhs, we get an energy efficiency ratio of 907 cal/kWh. We're looking for high calories divided by low energy. So a higher energy efficiency ratio is better than low. Here's some different foods and their ratios:

Energy Efficiency Ratio (cal/KWh)
per Pound

----

907 – Corn
363 – Milk (cow)
253 – Cheese
168 – Eggs
129 – Apples
102 – Chicken
52 – Pork
37 – Beef

As you can see, calories from corn are 25 times more energy efficient than beef. Why is beef always on the wrong end of things?

Bottom line: We can feed twice as many people using a fraction of the energy if we eat the crop instead of the cow.

Food And Health

There are lots of studies linking meat and dairy to increased rates of chronic disease. Things like heart disease, stroke, diabetes, cancer, and other diseases automatically associated with a western diet. You know, that "western diet". The one with all of those pre-packaged foods, refined grains, red meats, fried foods, sodas, candies, fatty dairy products, etc., etc. Not to mention the general rejection of healthy fruits, vegetables, whole grains, fish, nuts, and seeds.

The World Health Organization says, "A healthy diet helps to protect against malnutrition in all its forms, as well as noncommunicable diseases (NCDs), including diabetes, heart disease, stroke and cancer."[107] The Center for Disease Control and Prevention (CDC) says, "Adults who eat a healthy diet live longer and have a lower risk of obesity, heart disease, type 2 diabetes, and certain cancers."[108] The National Institutes of Health (NIH) says, "Healthy eating can lower your risk for heart disease, stroke, diabetes, and other health conditions."[109]

A joint project by Cornell University, Oxford University and the Chinese Academy of Preventive Medicine, examined 65 Chinese counties to study the link between animal products and chronic illnesses like heart disease, diabetes, and cancers (breast, prostate, and bowel). They found that counties with higher consumption of animal-based foods had higher death rates, while counties with a higher consumption of plant-based foods had lower death rates. The researchers concluded that people who eat a predominantly whole-food, plant-based diet will "escape, reduce, or reverse the development of chronic diseases." The results were published in *The China Study.*[110]

Another study at the Cleveland Clinic used a plant-based diet to treat patients with advanced heart disease. After five years, the surprising results were the general reduction of patient cholesterol levels by 50%, a halting of disease progression in ALL participants and the actual reversal of the disease in 44% of the patients. In short, these patients were ALL able to return to a normal, active life. This study was published in the American Journal of Cardiology.[111]

Colon cancer is the second leading cause of cancer deaths in the United States. African Americans are 100 times more likely to develop colon cancer than South Africans. A study swapped the diets of these two groups. The African Americans ate a typical rural African diet of beans and vegetables, and the South Africans ate a western diet high in fat and animal protein. After just two weeks, the African Americans showed a significant drop in cancer markers and the South Africans showed a significant increase.[112]

We've known for decades that a plant-rich diet has amazing health benefits, but the U.S. and other western countries remain chronically stuck in a processed food, red meat diet. Where's the beef!

Bottom line: The foods that feed chronic diseases are the same foods driving that 24% of global greenhouse gas emissions, are the same foods consuming 59% of all available land and way too much of our fresh water supplies, and the same foods that cost us the most at the grocers. Red meats and processed foods – that is, the western diet.

Food And Longevity

National Geographic did a cover story in 2005 called "The Secrets of a Long Life". They looked at world age statistics, and identified five regions of the world with the highest concentration of centenarians – people over 100 years old. They circled these regions on a map with a blue marker and started calling them the Blue Zones.[113] They found that people living in these blue zones had several things in common. They didn't smoke. Their lives involved some form of regular moderate physical activity. And they ate a plant-based diet, heavy in beans, peas, and lentils – that is, legumes.

Bottom line: Dried legumes are super cheap, have a shockingly long shelf life, and are one of the most nutritious foods people can eat. They're also super rich in protein, making them a great substitute for meat.

Food And Economy

Loss of income that comes from being ill can add up. Loss of income from being chronically ill can be crushing. Even with insurance, medical and drug expenses can wipe a family out when chronic disease strikes. A diet that prevents disease lets you earn more money and pay less for healthcare. When we talk about prospering in a post-climate change world, we're talking about this. Eating in a way that is nutritious, healthy, saves you money and slows the rate of climate change. Now, that's a prosperous life.

Remember that study mentioned earlier that showed how different dietary styles reduced greenhouse gas emissions? That study also looked at health issues and work productivity. It found that moving the world to that doctor-recommended healthy global diet (HGD) would prevent over five million deaths per year. Moving to a vegetarian diet would prevent over 7 million deaths and moving everyone to a vegan diet would prevent over 8 million deaths per year. The study also looked at the overall cost-of-illness, both direct healthcare expenses and lost work income. It found that the HGD would save $735 billion per year, the vegetarian diet would save $973 billion and the vegan diet would save $1.06 trillion per year. Big numbers, but not as big as the study's value-of-life projections. These projections are the values assigned by governments and insurance companies to a lifetime of work. The study found that the HGD would lead to an increase value-of-life of $21 trillion per year, the vegetarian diet $28 trillion and the vegan diet $30 trillion per year.

Lives Saved by Dietary Style
per Year

----

5 million – HGD
7 million – Vegetarian
8 million – Vegan

Cost-of-Illness Saved by Dietary Style
per Year

$735 billion – HGD
$973 billion – Vegetarian
$1.06 trillion – Vegan

Value-of-Lives Saved by Dietary Style per Year

$21 trillion – HGD
$28 trillion – Vegetarian
$30 trillion – Vegan

Bottom line: What you put on your plate really does matter. The less red meat and processed food you eat, the better for basically everything – your health, your pocketbook, the economy and the planet.

But I Need Protein

Yes you do. All animals need protein. Our bodies are made of protein. And proteins are made up of amino acids. Amino acids are considered the building blocks of life.[114] Remember photosynthesis? Plants scrubbing that CO_2 from the air? Well, plants take that CO_2 scrubbed from the air, break it down into carbon & oxygen, then combine it with hydrogen & nitrogen to create amino acids. That's where amino acids come from. Plants. When we eat a steak, we're getting protein because that cow ate a lot of grass and used the grass' amino acids to create cow protein. We eat the cow, break the cow protein back down into amino acids, then put those amino acids back together into people protein. Wouldn't it be easier if we could just ate the plants?

> *TRIVIA: There are 20 different amino acids and all plants can make all 20 of them. Animals can make 11 of the amino acids, but not the other nine. These nine are called the essential amino acids. Animals have to get them from plants, either by eating the plant or by eating the animal that ate the plant.*

Every plant contains every amino acid. The problem is that

different plants have different amounts of each amino. And sometimes we'd have to eat an awful lot of a plant to get enough of all the essential amino acids. For instance, we could eat five potatoes and get the full RDI (Recommended Daily Intake) of all essential amino acids. But who wants to eat five potatoes? It's just easier to eat the cow, which has already eaten enough grass to have all the amino acids we need.

Of course, eating cow is bad for our health and creates lots of greenhouse gases. Fortunately, it turns out that there are plants that have even more amino acids than cows. For instance, plants like quinoa, buckwheat, tofu, tempeh, edamame, soymilk, nutritional yeast, pistachios, hempseed, chia, pumpkin seeds, etc. have all of the essential amino acids needed for our bodies to build complete people proteins without ever cooking a cow.

And other plants can be eaten in combination to give us a complete set of essential amino acids. There are four groups of amino acid-related plants – grains, legumes, veggies and nuts/seeds – and eating any two of these groups together will give you all nine essential amino acids. For instance, legumes are low in the amino acids methionine and tryptophan but have plenty of lysine and isoleucine. Nuts/seeds are low in the amino acids lysine and isoleucine but have plenty of methionine and tryptophan. So throw together some pumpkin seeds and peanuts and "voilà" – you've got everything your body needs to build its people proteins. The same goes for combinations like beans & rice, hummus & pita, bean-chili and crackers, refried beans and tortillas. There's a reason these food combinations are so popular. They're complete proteins. Even the common peanut butter sandwich is a complete protein combination of legumes (peanuts) and grains (bread).

The lowly black bean is a good example of plant-based protein. Let's compare it to beef.

Nutritional Value of Beef per Pound

1152 – Calories
91 g – Fat
0 g – Carbohydrates
78 g – Protein

Nutritional Value of Black Beans per Pound

1548 – Calories
19 g – Fat
283 g – Carbohydrates
98 g – Protein

As you can see, pound for pound, black beans have less fat and provide more calories and protein than beef. Just keep in mind that black beans are legumes and a little low in a couple of essential amino acids, so throw a grain like barley or corn into that black-bean chili.

Bottom line: Be plant-forward. In today's more climate-savvy world, there's a growing buffet of really good plant-based replacements for meats, eggs, cheese and dairy. It's easier than ever to escape from the "western diet". And you can be plant-forward without totally giving up meat. Just think about the plants you put on your plate first, then consider some meat on the side. Cut back hard on the red meats and enjoy other proteins. Gradually introduce more of those great, savory plant-based soups, curries and stews into your diet.

* * *

Talking Points

Where's the beef? It's over there creating climate change!

You can grow 37,200 pounds of corn on one acre, or you can grow 354 pounds of beef. What if we ate the corn instead of the cow?

83% of all farmland is used to raise livestock that provides 18% of the world's calories.

17% of all farmland is used to raise food crops that provide 82% of the world's calories.

It takes around 1100 gallons of water to make one gallon of cow milk.

The western diet causes chronic diseases like heart disease, diabetes, and cancer.

A plant-based diet can "escape, reduce, or reverse the development of chronic diseases."

Cows get all their protein from grass.

Black beans contain more protein than beef.

If growing meat in the lab is the same as real meat, will it be just as bad for our health?

Action Items

Eat less red meat.

Try some of the amazing new plant-based meat, eggs, cheese and dairy.

Pick one day a week to go meatless.

Be plant-forward. Plan the plants on your plate first.

WHAT WE WEAR

PRODUCTS – The Industrial Sector

The industrial sector emits 21% of global greenhouse gases. That's right up there with power production and agriculture. Unless you own a factory, you probably don't have much control over manufacturing. What you DO have control over is the products you buy, use and wear. You may not think that your purchases can make a difference, but think again.

There are 74 recognized industries[115] and one of them – just ONE of them – is responsible for 48% of ALL INDUSTRIAL greenhouse gas emissions worldwide. That's more than the total greenhouse gas emissions of the international airline and shipping industries combined. That's more than the total greenhouse gas emissions of the entire country of Russia.

What we're talking about is the clothing on your back. The fashion industry accounts for 10% of all global greenhouse gas emissions.[116] And much of that is due to the type of textile used in your clothing.

Textiles

Textile is the fiber or filament used to create fabrics. Twisted together, they become the thread or yarn that is woven into cloth. Textiles are the materials used to make your clothing, curtains, carpets, rugs, upholstery and anything else that you look at and say, "Hey, that's made out of cloth." There are natural textile fibers made from plants. And then there are man-made textile filaments.

The thing that may surprise you is that most man-made textiles are petro-textiles. Petro, as in made out of petroleum. That's right. The same gooey crude oil used to make gasoline, diesel, and kerosene. This crude oil is refined into plastic that is "extruded" into endless filaments and woven into fabrics. You're probably wearing fossil fuel right now. And the same greenhouse gas emissions we'd expect from all the drilling, refining, shipping and burning of fossil fuels, apply to your clothing.

The energy required to produce various materials is a good measure of their greenhouse gas emissions. Here's a list of textiles and the energy it takes to produce them.[117] For reference, let's throw in some common industrial products, like glass and steel.

MegaJoules (MJ) Energy Needed to Produce Product per Kilogram

----

10 MJ – Linen
13 MJ – Hemp
15 MJ – Glass
36 MJ – Steel
60 MJ – Cotton
71 MJ – Rayon
125 MJ – Polyester
175 MJ – Acrylic
250 MJ – Nylon

Wow! Can you believe that it actually takes 3-6 times more energy to produce petro-textiles than it does to produce industrial glass and steel? It's no surprise that our natural, plant-based textiles fall well below these levels.

Over 63% of all textiles are now petroleum-based. Eighty percent of that is polyester. Around 60 million tonnes of polyester was made in 2021.[118] And making polyester releases over 1.5 trillion pounds of greenhouse gases each year. That's about the same emissions as 185 coal-fired power plants.[119]

> *TIP: Of the plant-based textiles, linen and hemp are best. They use significantly less energy than cotton and rayon (which is made out of plant cellulose). And traditionally farmed cotton uses way too much water, herbicides and pesticides to be good for any of us.*

Bottom line: Petro-textiles – things like polyester, acrylic, nylon, spandex and lycra – are made out of petroleum. If we're taking the fossil out of fuel, we also need to get the fossil out of fabric.

Fast Fashion

Petro-textiles are bad for the environment. But the reason the fashion industry has the catastrophic potential to consume more than 26% of the world's carbon budget by 2050, is that we've super-charged the rate at which we purchase and dispose of garments.[120] We used to get new clothing twice a year. There would be the spring fashion line and the fall fashion line. Very exciting. This changed In the 1990s, when companies started introducing new fashion lines every few weeks. Zara, a big fashion retailer in Europe, puts out around 24 collections a year, and other fashion retailers, like H&M, put out 12 to 16.[121] The world has moved from two fashion cycles a year to fifty.

The Merriam Webster dictionary defines Fast Fashion as, "an approach to the design, creation, and marketing of clothing fashions that emphasizes making fashion trends quickly and cheaply available to consumers."

Investopedia says, "Fast Fashion is the term used to describe clothing designs that move quickly from the catwalk to stores. In fact, it's not uncommon for fast-fashion retailers to introduce new products multiple times in one week to stay on trend."[122]

This business model is based on selling clothes cheaply and frequently. It's cheap, so you buy more often. You're buying more often so it doesn't need to last as long. It doesn't need to last as long so it can be constructed cheaply. It's cheap, so you can buy more often. You're buying more often so the cycle of consumerism and manufacturing just continues to grow.

Fast Fashion makes shopping for clothes more affordable, but it comes at a steep environmental cost.[123] People are buying 60% more clothing and keeping it half as long. Discarded clothing ends up in the landfills or being burned. The EPA reports that 34 million pounds of clothing were discarded in 2018.[124] Around five million pounds of that was recycled, six million pounds were burned, and the remaining 23 million pounds were buried in landfills.

Not good. Burning petro-textiles has the same effect as burning any other fossil fuel. And landfills create their own environmental problems that we'll discuss later.

Bottom line: Fast fashion accelerates consumerism and manufacturing, leading to over-consumption and resource depletion. Big words for taking too much and not leaving enough behind.

But It Saves Me Money

Does it really? You're buying a fast fashion shirt every couple of weeks, wearing it twice, then the fabric tears or buttons come off and you throw it away. If you're paying $16 for that shirt, you've payed $8 per use. If instead, you bought a good quality shirt for $100, then wore it once a month for a couple years, you'd be paying closer to $4 per use.

Bottom line: You actually save money by investing in good

clothing.

What About My Health

The fashion industry produces 20% of global industrial water pollution.[125] Once again, that's ONE industry out of 74 producing more than its share of trouble. And there are around 8,000 synthetic chemicals and 2,400 different substances used in clothing manufacturing, 30% of which may be toxic, cancerous, or pose some other risk to human health.

Fast Fashion is built on cheap petro-textiles like polyester, acrylic, nylon, spandex and lycra. These are all different types of plastic and major contributors to the microfiber and microplastic health crisis. Research at the University of Plymouth[126] found that a single load of fast fashion laundry can release over 700,000 microfibers. These tiny fibers make their way into our water, where they clump together, become home to colonies of pathogenic bacteria[127] and then work their way up our food chain.

And fast fashion slowly breaks down to form microplastics. Microplastics are now in our air, our water, our food and probably everyplace else. Microplastics have been found in our lungs,[128] in breast milk,[129] in placentas,[130] in blood[131] and throughout our digestive systems. A bottle-fed baby swallows millions of micoplastic pieces each day.[132]

Bottom line: Researchers are just beginning to study the health impacts of the plastics pandemic. Be safe and limit your exposure. Learn more in the upcoming *Field Guide to Plastics.*

Landfills And Donations

The world dumps over 90 million tons of fast fashion into landfills each year. Petro-textiles take centuries to break down, so those landfills will be around long after we're gone. A thousand years from now, archeologists will be defining this era by the plastic clothing we've buried.

A lot of us think that donating our clothing helps. Frankly, it doesn't. Only 10% of donated clothing reaches the thrift store racks. The rest is sold to textile recyclers who make rags out of what they can, then bale up the rest and ship it to landfills in other countries. We really shouldn't even call them landfills. They're basically mountains of clothing being dumped in places like Chile's Atacama desert.[133] Locals may scavenge for a few fast fashion designer pieces to sell in local flea markets, but that's not even a decimal point to the thousands of tons of clothing dumped at each site annually – specifically 39,000 tons a year in the case of the Atacama desert.

Bottom line: Donating is not the answer. The answer is buying less and wearing it longer.

Buy Less And Wear It Longer

Changing what you wear and the way you shop can save you money, create a safer environment and help slow climate change. What's not to like? And the changes are pretty simple.

One of the simplest is to just stop buying clothes made out of fossil fuel. Polyester, acrylic, nylon, spandex, lycra ... they're all petro-textiles. Read the garment labels and just say no. Buy plant-based fabrics like linen and hemp instead. They're comfortable, durable, healthy and great for the environment. If you want to buy cotton, look for organic. It will reduce your exposure to the residual herbicides and pesticides found on traditional cotton.

And change the way you shop for clothing. It will save you lots of money and create a healthier, safer world. Be conscious and mindful when you shop. Shop for clothing that is timeless, not trendy. Think of clothing as a long-term investment instead of a short-term fast fashion. Focus on quality instead of quantity. Think sustainable and don't view clothing as disposable.

Shop less frequently and shop your closet. How many things are buried there that you've forgotten? Shop secondhand and vintage. Older clothing is more durable than modern fast fashion. Instead of buying something, swap clothing with friends and family. Rent your glamour, your tuxes and gowns. Avoid the fast fashion brands and shop small local outlets. If the place you're shopping only sells petro-textiles, you're in the wrong store.

Bottom line: Don't let the fashion industry dictate what you wear. Own your fashion style. Repair and repurpose clothing instead of throwing it away. Launder less & hang dry to make your clothing last longer.

* * *

Talking Points

Are you wearing refined crude oil? Check the labels.

A polyester-cotton blend is like a hybrid car. It's better than 100% fossil fuel but it's not as good as zero.

Linen and hemp fabrics are comfortable, durable, and great for the environment.

A plastic microfiber is less than 10 microns thick. That's REALLY small.

A single load of laundry can release over 700,000 plastic microfibers into our rivers and oceans.

There are an estimated 1.4 quadrillion microfibers in the ocean. That's 1.4 followed by 15 zeros.

Action Items

Stop buying polyester, nylon, acrylic, spandex and lycra.

Buy less and wear it longer.

Learn how to sew on a button. There are plenty of how-to videos online.

If you need a piece of clothing, shop your closet before heading to the second-hand store.

Shop your friends and families' closets.

If you've never had one, buy a linen or hemp shirt.

ABOUT THE AUTHOR

Bob Grove grew up among northern Missouri farmers where he developed a passion for the rights of people and for healthy environments in which those rights can thrive. He currently serves as President of the Climate Council (www.ClimateGKC.org) and Executive Producer of the Climate Hour (www.ClimateHour.net). Bob is a frequent speaker on a wide range of sustainability topics, both in person and through online workshops. He can be reached through Wordsinger Publishing (www.Wordsinger.com). Watch for Bob Grove's next book, *Grove's Field Guide to Plastics*, coming in 2024.

BIBLIOGRAPHY

Because internet content can change over time, we do not guarantee that the URLs provided here will continue to work.

Airbus. “Hybrid and Electric Flight.” Accessed March 27, 2023. http://www.airbus.com/en/innovation/zero-emission-journey/hybrid-and-electric-flight.

Ali, Shirin. "Massive Mountain of Clothing in the Desert Is Growing Even Bigger." The Hill, November 8, 2021. Accessed March 25, 2023. http://thehill.com/changing-america/sustainability/environment/580521-massive-mountain-of-clothing-in-the-desert-is.

Archer Aviation. "United Airlines And Archer Announce First Commercial Electric Air Taxi Route In Chicago," March 23, 2023. Accessed March 25, 2023. http://www.archer.com/news/united-airlines-and-archer-announce-first-commercial-electric-air-taxi-route-in-chicago.

Associated Press. "Hawaiian Airlines Exploring Electric Airplane Technology." US News & World Report, May 12, 2022. Accessed March 25, 2023. http://www.usnews.com/news/best-states/hawaii/articles/2022-05-12/hawaiian-airlines-exploring-electric-airplane-technology.

Baker, Thomas. "Read This Before You Buy Ductless AC." This Old House, March 13, 2023. Accessed March 25, 2023. https://www.thisoldhouse.com/heating-cooling/21018989/read-this-before-you-buy-ductless-ac.

Barrett, John, Matthew Chadwick, and Mike Chadwick.

"Ecological Footprint and Water Analysis of Cotton, Hemp and Polyester." Stockholm Environment Institute, March 11, 2020. Accessed March 25, 2023. https://www.sei.org/publications/ecological-footprint-water-analysis-cotton-hemp-polyester.

Batten, Robert, Jennifer L Turner, and Susan Chan Shifflett. "China's Upstream Advantage in the Great Himalayan Watershed - The National Bureau of Asian Research (NBR)." The National Bureau of Asian Research (NBR), July 15, 2013. Accessed March 25, 2023. https://www.nbr.org/publication/chinas-upstream-advantage-in-the-great-himalayan-watershed.

Bland, Alastair. "Is the Livestock Industry Destroying the Planet?" Smithsonian Magazine, August 1, 2012. Accessed March 25, 2023. https://www.smithsonianmag.com/travel/is-the-livestock-industry-destroying-the-planet-11308007.

Blue Zones. "History of Blue Zones," April 30, 2021. Accessed March 25, 2023. https://www.bluezones.com/about/history.

Bryan, Victoria. "Hawaiian Airlines Invests in Electric Seaglider Maker for Inter-Island Flights -AeroTime." AeroTime, May 11, 2022. Accessed March 25, 2023. https://www.aerotime.aero/articles/30983-hawaiian-airlines-electric-seaglider-regent.

Buis, Alan. "Why Milankovitch (Orbital) Cycles Can't Explain Earth's Current Warming." NASA Climate Change: Vital Signs of the Planet, February 27, 2022. Accessed March 25, 2023. https://climate.nasa.gov/ask-nasa-climate/2949/why-milankovitch-orbital-cycles-cant-explain-earths-current-warming.

Bustos, Laura Williams, M.S. "What Is the Average Cycling Speed?" Bicycling, February 28, 2023. Accessed

March 25, 2023. https://www.bicycling.com/training/a43031175/average-cycling-speed.

Campbell, T. Colin, & Thomas M. Campbell. The China Study: The Most Comprehensive Study of Nutrition Ever Conducted and the Startling Implications for Diet, Weight Loss and Long-Term Health. BenBella Books, 2006.

Carrington, Damian. "Avoiding Meat and Dairy Is 'Single Biggest Way' to Reduce Your Impact on Earth." The Guardian, May 31, 2018. Accessed March 25, 2023. https://www.theguardian.com/environment/2018/may/31/avoiding-meat-and-dairy-is-single-biggest-way-to-reduce-your-impact-on-earth.

Carrington, Damian. "Bottle-Fed Babies Swallow Millions of Microplastics a Day, Study Finds." The Guardian, October 20, 2020. Accessed March 25, 2023. https://www.theguardian.com/environment/2020/oct/19/bottle-fed-babies-swallow-millions-microplastics-day-study.

Carrington, Damian. "Humans Just 0.01% of All Life but Have Destroyed 83% of Wild Mammals – Study." The Guardian, May 21, 2018. Accessed March 25, 2023. https://www.theguardian.com/environment/2018/may/21/human-race-just-001-of-all-life-buthas-destroyed-over-80-of-wild-mammals-study.

Carrington, Damian. "Microplastics Found in Human Breast Milk for the First Time." The Guardian, October 7, 2022. Accessed March 25, 2023. https://www.theguardian.com/environment/2022/oct/07/microplastics-human-breast-milk-first-time.

Centers for Disease Control and Prevention (CDC). "Poor Nutrition." Accessed March 25, 2023. https://www.cdc.gov/chronicdisease/resources/publications/factsheets/nutrition.htm.

Cleveland Clinic. "What Are Complete Proteins?," December 6, 2022. Accessed March 25, 2023. https://health.clevelandclinic.org/do-i-need-to-worry-about-eating-complete-proteins.

Climate Action Tracker. "Temperatures," November 11, 2022. Accessed March 25, 2023. https://climateactiontracker.org/global/temperatures.

Domonoske, Camila. "General Motors Sets All-Electric Target For Vehicles By 2035." NPR, February 1, 2021. Accessed March 25, 2023. https://www.npr.org/2021/02/01/962946561/general-motors-sets-all-electric-target-for-vehicles-by-2035.

Dove, Laurie L. "Is Induction Cooking Better Than Gas or Electric?" HowStuffWorks, March 8, 2023. Accessed March 25, 2023. https://recipes.howstuffworks.com/tools-and-techniques/induction-cooking.htm?srch_tag=3bvz23ntle7fhn4p74izqkdmqncjur2i.

Drew, Deborah, and Genevieve Yehounme. "The Apparel Industry's Environmental Impact in 6 Graphics." World Resources Institute, July 5, 2017. Accessed March 25, 2023. https://www.wri.org/insights/apparel-industrys-environmental-impact-6-graphics.

Eartheasy. "LED Light Bulbs: Comparison Charts." Eartheasy Guides & Articles, February 15, 2022. Accessed March 25, 2023. https://learn.eartheasy.com/guides/led-light-bulbs-comparison-charts.

Ellen MacArther Foundation. "Fashion and the Circular Economy." Accessed March 25, 2023. https://ellenmacarthurfoundation.org/fashion-and-the-circular-economy-deep-dive.

Encyclopedia Britannica. "Industrial Revolution | Definition,

History, Dates, Summary, & Facts," July 20, 1998. Accessed March 25, 2023. https://www.britannica.com/event/Industrial-Revolution.

Energy Star. "About ENERGY STAR." Accessed March 25, 2023. https://www.energystar.gov.

Energy Star. "ENERGY STAR Certified Heat Pumps (Ducted)." Accessed March 25, 2023. https://www.energystar.gov/productfinder/product/certified-central-heat-pumps.

Energy Star. "ENERGY STAR Certified Heat Pumps (Mini-Split)." Accessed March 25, 2023. https://www.energystar.gov/productfinder/product/certified-mini-split-heat-pumps.

Energy Star. "Kitchen Appliances." Accessed March 25, 2023. https://www.energy.gov/energysaver/kitchen-appliances.

Esselstyn, Caldwell B. "Updating a 12-Year Experience with Arrest and Reversal Therapy for Coronary Heart Disease (an Overdue Requiem for Palliative Cardiology)." American Journal of Cardiology 84, no. 3 (August 1, 1999): 339–41. https://doi.org/10.1016/s0002-9149(99)00290-8.

Evans, Simon. "Solar, Wind and Nuclear Have 'Amazingly Low' Carbon Footprints, Study Finds." Carbon Brief, December 8, 2017. Accessed March 25, 2023. https://www.carbonbrief.org/solar-wind-nuclear-amazingly-low-carbon-footprints.

Franta, Benjamin. "On Its 100th Birthday in 1959, Edward Teller Warned the Oil Industry about Global Warming." The Guardian, January 1, 2018. Accessed March 25, 2023. https://www.theguardian.com/environment/climate-consensus-97-per-cent/2018/jan/01/on-its-hundredth-birthday-in-1959-edward-teller-warned-the-oil-industry-about-global-warming.

Go Climate. "Travel Emissions Calculator." Accessed March 25,

2023. https://www.goclimate.com/us/travel-calculator.

Gutierrez, Miren, and Guy Jobbins. "China's Distant-Water Fishing Fleet: Scale, Impact and Governance." ODI, June 2, 2020. Accessed March 25, 2023. https://odi.org/en/publications/chinas-distant-water-fishing-fleet-scale-impact-and-governance.

Guzman, Francisco. "How LED Light Bulbs Work." HowStuffWorks, March 8, 2023. Accessed March 25, 2023. https://science.howstuffworks.com/environmental/green-tech/sustainable/led-light-bulb.htm.

Hall, Shannon. "Exxon Knew about Climate Change Almost 40 Years Ago." Scientific American, October 26, 2015. Accessed March 25, 2023. https://www.scientificamerican.com/article/exxon-knew-about-climate-change-almost-40-years-ago.

Harris, William. "How CFL Bulbs Work." HowStuffWorks, March 8, 2023. Accessed March 25, 2023. https://science.howstuffworks.com/environmental/green-tech/sustainable/cfl-bulb.htm.

Harvard T.H. Chan School of Public Health. "Home Is Where the Pipeline Ends," June 28, 2022. Accessed March 25, 2023. https://www.hsph.harvard.edu/c-change/news/home-is-where-the-pipeline-ends.

Harvard T.H. Chan School of Public Health. "Natural Gas Used in Homes Contains Hazardous Air Pollutants," June 28, 2022. Accessed March 25, 2023. https://www.hsph.harvard.edu/c-change/news/natural-gas-used-in-homes.

Hawkin, Paul. "Drawdown." Resilience, June 21, 2017. Accessed March 25, 2023. https://www.resilience.org/resources/drawdown.

Hayes, Adam. "Fast Fashion Explained and How It Impacts Retail Manufacturing." Investopedia, September 16, 2022.

Accessed March 25, 2023. https://www.investopedia.com/terms/f/fast-fashion.asp.

HowStuffWorks. "How Does a Halogen Light Bulb Work?," March 8, 2023. Accessed March 25, 2023. https://home.howstuffworks.com/question151.htm.

Jordan, Rob. "Rethinking Cooking with Gas." Stanford News, January 27, 2022. Accessed March 25, 2023. https://news.stanford.edu/2022/01/27/rethinking-cooking-gas.

Kerlin, Kat. "The Diet That Helps Fight Climate Change." UC Davis, December 28, 2017. Accessed March 25, 2023. https://www.ucdavis.edu/news/diet-helps-fight-climate-change.

Ladyzhets, Betsy. "How Much Does Eating Meat Affect Nations' Greenhouse Gas Emissions?" Science News, May 5, 2022. Accessed March 25, 2023. https://www.sciencenews.org/article/food-emissions-data-diet-carbon-greenhouse-gas-climate-agriculture.

Laybourn, Laurie, Henry Throp, and Suzannah Sherman. "1.5°C – Dead or Alive? The Risks to Transformational Change from Reaching and Breaching the Paris Agreement Goal." Institute for Public Policy Research, February 16, 2023. Accessed March 25, 2023. https://www.ippr.org/research/publications/1-5c-dead-or-alive.

Leslie, H. M., Martin Van Velzen, Sicco H. Brandsma, Dick Vethaak, Juan J. Garcia-Vallejo, & Marja H. Lamoree. "Discovery and Quantification of Plastic Particle Pollution in Human Blood." Environment International 163 (March 1, 2022): 107199. https://doi.org/10.1016/j.envint.2022.107199.

Li, Dunzhu, Yunhong Shi, Luming Yang, Liwen Xiao, Daniel K. Kehoe, Yurii K. Gun'ko, John J. Boland, and Jing Wang. "Microplastic Release from the Degradation of Polypropylene Feeding Bottles during Infant Formula Preparation." Nature Food 1, no. 11 (November 1, 2020):

746–54. https://doi.org/10.1038/s43016-020-00171-y.

Lindsey, Rebecca. "Climate Change: Atmospheric Carbon Dioxide." NOAA Climate.gov, June 23, 2022. Accessed March 25, 2023. https://www.climate.gov/news-features/understanding-climate/climate-change-atmospheric-carbon-dioxide.

Maron, Dina Fine. "WHO Moves to Contain Superbugs on the Farm." Scientific American, November 7, 2017. Accessed March 25, 2023. https://www.scientificamerican.com/article/who-moves-to-contain-superbugs-on-the-farm.

Matthews, M. A. and Shell International Chemical Company Ltd. "The Earth's Carbon Cycle." The New Scientist, October 8, 1959, 644–46. Accessed March 25, 2023. https://www.documentcloud.org/documents/4447739-The-Earth-s-Carbon-Cycle-Shell.

McFall-Johnsen, Morgan. "These Facts Show How Unsustainable the Fashion Industry Is." World Economic Forum, January 31, 2020. Accessed March 25, 2023. https://www.weforum.org/agenda/2020/01/fashion-industry-carbon-unsustainable-environment-pollution.

Michanowicz, Drew R., Archana Dayalu, Curtis L. Nordgaard, Jonathan J. Buonocore, Molly Wilson Fairchild, Robert C. Ackley, Jessica E Schiff, et al. "Home Is Where the Pipeline Ends: Characterization of Volatile Organic Compounds Present in Natural Gas at the Point of the Residential End User." Environmental Science & Technology 56, no. 14 (June 28, 2022): 10258–68. https://doi.org/10.1021/acs.est.1c08298.

Millennium Alliance for Humanity and the Biosphere (MAHB). "When Fossil Fuels Run Out, What Then?," May 23, 2019. Accessed March 25, 2023. https://mahb.stanford.edu/library-item/fossil-fuels-run.

Milman, Oliver. "Revealed: Exxon Made 'Breathtakingly' Accurate Climate Predictions in 1970s and 80s." The Guardian, January 16, 2023. Accessed March 25, 2023. https://www.theguardian.com/business/2023/jan/12/exxon-climate-change-global-warming-research.

Mobius, Ariel. "How Do Greenhouse Gases Trap Heat in the Atmosphere?." MIT Climate Portal, February 19, 2021. Accessed March 25, 2023. https://climate.mit.edu/ask-mit/how-do-greenhouse-gases-trap-heat-atmosphere.

Napper, Imogen E., and Richard F. Thompson. "Release of Synthetic Microplastic Plastic Fibres from Domestic Washing Machines: Effects of Fabric Type and Washing Conditions." Marine Pollution Bulletin 112, no. 1–2 (November 15, 2016): 39–45. https://doi.org/10.1016/j.marpolbul.2016.09.025.

NASA Earth Observatory. "Global Warming," June 3, 2010. Accessed March 25, 2023. https://earthobservatory.nasa.gov/features/GlobalWarming/page3.php.

NASA Global Climate Change: Vital Signs of the Planet. "Global Surface Temperature," March 7, 2023. Accessed March 25, 2023. https://climate.nasa.gov/vital-signs/global-temperature.

National Institutes of Health (NIH). "How Dietary Factors Influence Disease Risk," March 14, 2017. Accessed March 25, 2023. https://www.nih.gov/news-events/nih-research-matters/how-dietary-factors-influence-disease-risk.

National Oceanic and Atmospheric Administration. "What's the Difference between Climate and Weather?" Accessed March 25, 2023. https://www.noaa.gov/explainers/what-s-difference-between-climate-and-weather.

New England Vegetable Guide. "Table 15: Approximate Yields," May 26, 2022. Accessed March 25, 2023. https://nevegetable.org/cultural-practices/table-15-approximate-yields.

NOAA National Centers for Environmental Information. "Billion-Dollar Weather and Climate Disasters | National Centers for Environmental Information (NCEI)," April 10, 2023. Accessed March 25, 2023. https://www.ncei.noaa.gov/access/billions.

OEcotextiles. "What Is the Energy Profile of the Textile Industry?," October 31, 2011. Accessed March 25, 2023. https://oecotextiles.blog/2009/06/16/what-is-the-energy-profile-of-the-textile-industry.

O'Keefe, Sean F., Jia Li, Leo Lahti, Junhai Ou, Franck Carbonero, Khaled Mohammed, Joram M. Posma, et al. "Fat, Fibre and Cancer Risk in African Americans and Rural Africans." Nature Communications 6, no. 1 (April 28, 2015). https://doi.org/10.1038/ncomms7342.

OpenSecrets. "Oil & Gas Lobbying Profile." Accessed March 25, 2023. https://www.opensecrets.org/federal-lobbying/industries/summary?cycle=2022&id=E01.

OpenSecrets. "Oil & Gas." Accessed March 25, 2023. https://www.opensecrets.org/industries/indus.php?ind=E01.

Pattee, Emma. "The Scientists Hired by Big Oil Who Predicted the Climate Crisis Long Ago." The Guardian, June 13, 2022. Accessed March 25, 2023. https://www.theguardian.com/environment/2021/jul/02/scientists-climate-crisis-big-oil-climate-crimes.

Pedrotti, Maria Luiza, Ana Luzia De Figueiredo Lacerda, Stephanie Petit, Jean François Ghiglione, and Gabriel Gorsky. "Vibrio Spp and Other Potential Pathogenic Bacteria Associated to Microfibers in the North-Western Mediterranean Sea." PLOS

ONE 17, no. 11 (November 30, 2022): e0275284. https://doi.org/10.1371/journal.pone.0275284.

Peters, Christian J., Jamie Picardy, Amelia Darrouzet-Nardi, Jennifer L. Wilkins, Timothy J. Griffin, and Gary W. Fick. "Carrying Capacity of U.S. Agricultural Land: Ten Diet Scenarios." Elementa 4 (July 22, 2016). https://doi.org/10.12952/journal.elementa.000116.

Phys.org. "Microfibers in the Mediterranean Sea Are Floating Homes for Bacteria," November 30, 2022. Accessed March 25, 2023. https://phys.org/news/2022-11-microfibers-mediterranean-sea-homes-bacteria.html.

Polakovic, Gary. "Water Dispute on the Nile River Could Destabilize the Region." USC News, July 13, 2021. Accessed March 25, 2023. https://news.usc.edu/188414/nile-river-water-dispute-filling-dam-egypt-ethiopia-usc-study.

Poore, Joseph, and Thomas Nemecek. "Reducing Food's Environmental Impacts through Producers and Consumers." Science 360, no. 6392 (June 1, 2018): 987–92. https://doi.org/10.1126/science.aaq0216.

Project Drawdown. "Table of Solutions," August 16, 2021. Accessed March 25, 2023. https://www.drawdown.org/solutions/table-of-solutions.

Ragusa, Antonio, Alessandro Svelato, Criselda Santacroce, Piera Catalano, Simona Sabbatini, Oliana Carnevali, Fabrizio Papa, et al. "Plasticenta: First Evidence of Microplastics in Human Placenta." Environment International 146 (January 1, 2021): 106274. https://doi.org/10.1016/j.envint.2020.106274.

Rannard, Georgina. "ExxonMobil: Oil Giant Predicted Climate Change in 1970s - Scientists." BBC News, January 12, 2023. Accessed March 25, 2023. https://www.bbc.com/news/science-environment-64241994.

Riebeek, Holli. "Paleoclimatology: The Oxygen Balance." NASA Earth Observatory, May 6, 2005. Accessed March 25, 2023. https://earthobservatory.nasa.gov/features/Paleoclimatology_OxygenBalance.

Ritchie, Hannah, and Max Roser. "Water Use and Stress." Our World in Data, July 2018. Accessed March 25, 2023. https://ourworldindata.org/water-use-stress#share-of-freshwater-withdrawals-used-in-agriculture.

Ritchie, Hannah. "Half of the World's Habitable Land Is Used for Agriculture." Our World in Data, November 11, 2019. Accessed March 25, 2023. https://ourworldindata.org/global-land-for-agriculture.

Ritchie, Hannah. "Which Form of Transport Has the Smallest Carbon Footprint?" Our World in Data, October 13, 2020. Accessed March 25, 2023. https://ourworldindata.org/travel-carbon-footprint.

Roser, Max. "Why Did Renewables Become so Cheap so Fast?" Our World in Data, December 1, 2020. Accessed March 25, 2023. https://ourworldindata.org/cheap-renewables-growth.

Sas, Wojciech. "Cattle per Acre Calculator." OMNI Calculator, January 18, 2023. Accessed March 25, 2023. https://www.omnicalculator.com/biology/cattle-per-acre.

Schlesinger, William. "The Fabric for Our Lives." Duke University: Nicholas School of the Environment, November 26, 2018. Accessed March 25, 2023. https://blogs.nicholas.duke.edu/citizenscientist/the-fabric-for-our-lives.

Sivak, Michael, and Brandon Schoettle. "Relative Costs of Driving Electric and Gasoline Vehicles in the Individual U.S. States." National Academies: Transportation Research Board, January 2018. Accessed March 31, 2023. https://trid.trb.org/view/1508116.

Scripps Institution of Oceanography. "The Keeling Curve." Accessed March 25, 2023. https://keelingcurve.ucsd.edu.

Snyder, Robert. "Leukemia and Benzene." International Journal of Environmental Research and Public Health 9, no. 8 (August 14, 2012): 2875–93. https://doi.org/10.3390/ijerph9082875.

Springmann, Marco, H. Charles J. Godfray, Mike Rayner, and Peter Scarborough. "Analysis and Valuation of the Health and Climate Change Cobenefits of Dietary Change." Proceedings of the National Academy of Sciences of the United States of America 113, no. 15 (April 12, 2016): 4146–51. https://doi.org/10.1073/pnas.1523119113.

Supran, G., S. Rahmstorf, and N. Oreskes. "Assessing ExxonMobil's Global Warming Projections." Science 379, no. 6628 (January 13, 2023). https://doi.org/10.1126/science.abk0063.

Sutherland, Paige. "The Water Fight over the Shrinking Colorado River." BBC News, April 8, 2021. Accessed March 25, 2023. https://www.bbc.com/news/world-us-canada-56608180.

Tabuchi, Hiroko. "Exxon Scientists Predicted Global Warming, Even as Company Cast Doubts, Study Finds." The New York Times, January 13, 2023. Accessed March 25, 2023. https://www.nytimes.com/2023/01/12/climate/exxon-mobil-global-warming-climate-change.html.

Textile Exchange. "Polyester," January 3, 2023. Accessed March 25, 2023. https://textileexchange.org/polyester.

The State of Food Security and Nutrition in the World 2022." FAO EBooks, July 6, 2022. https://doi.org/10.4060/cc0639en.

Thomas, Ian. "United Airlines Is Aiming to Have Electric Planes Flying by 2030." CNBC, October 10, 2022. Accessed March 25, 2023. https://www.cnbc.com/2022/10/06/united-airlines-is-aiming-to-have-electric-planes-flying-by-2030.html.

Thompson, Dennis. "Gas Used in Homes Has Links to Cancer; Leaks Often Undetected." US News & World Report, July 1, 2022. Accessed March 25, 2023. https://www.usnews.com/news/health-news/articles/2022-07-01/gas-used-in-homes-has-linksto-cancer-leaks-often-undetected.

Torchinsky, Rina. "For the First Time, Researchers Find Microplastics Deep in the Lungs of Living People." NPR, April 6, 2022. Accessed March 25, 2023. https://www.npr.org/2022/04/06/1091246691/microplastics-found-in-human-lungs.

Trollman, Hana. "Here's How the Russian Invasion Has Impacted Ukraine Crop Exports." World Economic Forum, April 20, 2022. Accessed March 25, 2023. https://www.weforum.org/agenda/2022/04/ukraine-grain-shortages-russia-conflict.

True Cost. "List of Foods By Environmental Impact and Energy Efficiency," January 26, 2016. Accessed March 25, 2023. https://truecostblog.com/2010/02/24/list-of-foods-by-environmental-impact-and-energy-efficiency.

UN News. "Ozone Layer Recovery Is on Track, Due to Success of Montreal Protocol," January 10, 2023. Accessed March 25, 2023. https://news.un.org/en/story/2023/01/1132277.

Union of Concerned Scientists. "Cleaner Cars from Cradle to Grave," October 29, 2015. Accessed March 25, 2023. https://www.ucsusa.org/resources/cleaner-cars-cradle-grave.

Union of Concerned Scientists. "The Disinformation Playbook," October 10, 2017. Accessed March 25, 2023. https://www.ucsusa.org/resources/disinformation-playbook.

United Nations Climate Change. "Key Aspects of the Paris

Agreement." Accessed March 25, 2023. https://unfccc.int/most-requested/key-aspects-of-the-paris-agreement.

United Nations Climate Change. "The Glasgow Climate Pact – Key Outcomes from COP26." Accessed March 25, 2023. https://unfccc.int/process-and-meetings/the-paris-agreement/the-glasgow-climate-pact-key-outcomes-from-cop26.

US Department of Energy. "2023 Chevrolet Bolt EV." Accessed March 25, 2023. https://www.fueleconomy.gov/feg/Find.do?action=sbs&id=45751.

US Department of Energy. "All-Electric Vehicles." Accessed March 25, 2023. https://fueleconomy.gov/feg/evtech.shtml.

US Department of Energy. "Alternative Fuels Data Center: Charging Electric Vehicles at Home." Accessed March 25, 2023. https://afdc.energy.gov/fuels/electricity_charging_home.html.

US Department of Energy. "Ductless Mini-Split Air Conditioners." Accessed March 25, 2023. https://www.energy.gov/energysaver/ductless-mini-split-air-conditioners.

US Department of Energy. "FOTW #1190, June 14, 2021: Battery-Electric Vehicles Have Lower Scheduled Maintenance Costs than Other Light-Duty Vehicles," June 14, 2021. Accessed March 25, 2023. https://www.energy.gov/eere/vehicles/articles/fotw-1190-june-14-2021-battery-electric-vehicles-have-lower-scheduled.

US Department of Energy. "FOTW #1230, March 21, 2022: More than Half of All Daily Trips Were Less than Three Miles in 2021," March 21, 2022. Accessed March 25, 2023. https://www.energy.gov/eere/vehicles/articles/fotw-1230-march-21-2022-more-half-all-daily-trips-were-less-three-miles-2021.

US Department of Energy. "History of Air Conditioning," July 20, 2015. Accessed March 25, 2023. https://www.energy.gov/articles/history-air-conditioning.

US Department of Energy. "Laundry." Accessed March 25, 2023. https://www.energy.gov/energysaver/laundry.

US Department of Energy. "Tankless or Demand-Type Water Heaters." Accessed March 25, 2023. https://www.energy.gov/energysaver/tankless-or-demand-type-water-heaters.

US Department of Energy. "Update or Replace Windows." Accessed March 25, 2023. https://www.energy.gov/energysaver/update-or-replace-windows.

US Department of Energy. "Weatherization Assistance Program." Accessed March 25, 2023. https://www.energy.gov/scep/wap/weatherization-assistance-program.

US Energy Information Administration. "Efficiency Requirements for Residential Central AC and Heat Pumps to Rise in 2023," July 30, 2019. Accessed March 25, 2023. https://www.eia.gov/todayinenergy/detail.php?id=40232.

US EPA. "Global Greenhouse Gas Emissions Data," February 15, 2023. Accessed March 25, 2023. https://www.epa.gov/ghgemissions/global-greenhouse-gas-emissions-data.

US EPA. "International Treaties and Cooperation about the Protection of the Stratospheric Ozone Layer," August 30, 2021. Accessed March 25, 2023. https://www.epa.gov/ozone-layer-protection/international-treaties-and-cooperation-about-protection-stratospheric-ozone.

US EPA. "Overview of Greenhouse Gases," May 16, 2022. Accessed March 25, 2023. https://www.epa.gov/ghgemissions/overview-greenhouse-gases.

US EPA. "Ozone-Depleting Substances," June 1, 2022. Accessed March 25, 2023. https://www.epa.gov/ozone-layer-protection/ozone-depleting-substances.

US EPA. "Plug-in Electric Vehicle Charging," February 22, 2023. Accessed March 25, 2023. https://www.epa.gov/greenvehicles/plug-electric-vehicle-charging.

US EPA. "Should You Have the Air Ducts in Your Home Cleaned?," February 27, 2023. Accessed March 25, 2023. https://www.epa.gov/indoor-air-quality-iaq/should-you-have-air-ducts-your-home-cleaned.

US EPA. "Sources of Greenhouse Gas Emissions," August 5, 2022. Accessed March 25, 2023. https://www.epa.gov/ghgemissions/sources-greenhouse-gas-emissions.

US EPA. "Textiles: Material-Specific Data," December 3, 2022. Accessed March 25, 2023. https://www.epa.gov/facts-and-figures-about-materials-waste-and-recycling/textiles-material-specific-data.

US EPA. "Understanding Global Warming Potentials," May 5, 2022. Accessed March 25, 2023. https://www.epa.gov/ghgemissions/understanding-global-warming-potentials.

USDA Economic Research Service. "U.S. Food Imports," March 23, 2023. Accessed March 25, 2023. https://www.ers.usda.gov/data-products/u-s-food-imports.

Vaidyanathan, Gayathri and Climate Wire. "Political Football over Climate Change Rattles Windows of Ivory Tower." Scientific American, March 23, 2015. Accessed March 25, 2023. https://www.scientificamerican.com/article/political-football-over-climate-change-rattles-windows-of-ivory-tower.

Water Footprint Calculator. "How Much Water Does It Take to

Make a Hamburger?," July 15, 2022. Accessed March 25, 2023. https://www.watercalculator.org/footprint/what-is-the-water-footprint-of.

Water Footprint of Food Guide - Water Footprint Calculator. (2022, November 23). Water Footprint Calculator. https://www.watercalculator.org/water-footprint-of-food-guide/

Wikipedia. "Climate Change Denial," March 16, 2023. Accessed March 25, 2023. https://en.wikipedia.org/wiki/Climate_change_denial.

Wikipedia. "Global Climate Coalition," February 4, 2023. Accessed March 25, 2023. https://en.wikipedia.org/wiki/Global_Climate_Coalition.

Wikipedia. "Global Industry Classification Standard," March 21, 2023. Accessed March 25, 2023. https://en.wikipedia.org/wiki/Global_Industry_Classification_Standard.

Wikipedia. "Global Warming Controversy," March 1, 2023. Accessed March 25, 2023. https://en.wikipedia.org/wiki/Global_warming_controversy.

Wikipedia. "Global Warming Potential," March 18, 2023. Accessed March 25, 2023. https://en.wikipedia.org/wiki/Global_warming_potential.

Wikipedia. "Information Council for the Environment," January 4, 2022. Accessed March 25, 2023. https://en.wikipedia.org/wiki/Information_Council_for_the_Environment.

Wikipedia. "Ozone Depletion," March 17, 2023. Accessed March 25, 2023. https://en.wikipedia.org/wiki/Ozone_depletion.

Wikipedia. "Tobacco Industry Playbook," March 23, 2023. Accessed March 25, 2023. https://en.wikipedia.org/wiki/Tobacco_industry_playbook.

Williams, Alan. "Washing Clothes Releases Thousands of

Microplastic Particles into Environment, Study Shows." University of Plymouth, September 27, 2016. Accessed March 25, 2023. https://www.plymouth.ac.uk/news/washing-clothes-releases-thousands-of-microplastic-particles-into-environment-study-shows.

WMO Lead Centre Annual-to-Decadal Climate Prediction. "World Meteorological Organization." Accessed March 25, 2023. https://hadleyserver.metoffice.gov.uk/wmolc.

Woodard, Collin. "This Electric Bus Drives 1,100 Miles on a Single Charge." MotorTrend, September 20, 2017. Accessed March 25, 2023. https://www.motortrend.com/news/electric-bus-claimed-1100-mile-range.

World Health Organization. "Healthy Diet," April 29, 2020. Accessed March 25, 2023. https://www.who.int/news-room/fact-sheets/detail/healthy-diet.

Worldometer. "World Population Clock: 8 Billion People (LIVE, 2023)." Accessed March 25, 2023. https://www.worldometers.info/world-population.

Yoder, Kate. "They Derailed Climate Action for a Decade. And Bragged about It." Grist, April 15, 2022. Accessed March 25, 2023. https://grist.org/accountability/how-the-global-climate-coalition-derailed-climate-action.

YouTube [TED]. "Shubhendu Sharma: How to Grow a Tiny Forest Anywhere," September 4, 2014. Accessed March 25, 2023. https://www.youtube.com/watch?v=3BgPFIKCaOQ.

ENDNOTES

Because internet content can change over time, we do not guarantee that the URLs provided here will continue to work.

---

[1]. NOAA National Centers for Environmental Information. "Billion-Dollar Weather and Climate Disasters | National Centers for Environmental Information (NCEI)," April 10, 2023. Accessed March 25, 2023. https://www.ncei.noaa.gov/access/billions/.

[2]. National Oceanic and Atmospheric Administration. "What's the Difference between Climate and Weather?" Accessed March 25, 2023. https://www.noaa.gov/explainers/what-s-difference-between-climate-and-weather.

[3]. Riebeek, Holli. "Paleoclimatology: The Oxygen Balance." NASA Earth Observatory, May 6, 2005. Accessed March 25, 2023. https://earthobservatory.nasa.gov/features/Paleoclimatology_OxygenBalance.

[4]. NASA Earth Observatory. "Global Warming," June 3, 2010. Accessed March 25, 2023. https://earthobservatory.nasa.gov/features/GlobalWarming/page3.php.

[5]. United Nations Climate Change. "Key Aspects of the Paris Agreement." Accessed March 25, 2023. https://unfccc.int/most-requested/key-aspects-of-the-paris-agreement.

[6]. United Nations Climate Change. "The Glasgow Climate Pact – Key Outcomes from COP26." Accessed March 25, 2023. https://unfccc.int/process-and-meetings/the-paris-agreement/the-glasgow-climate-pact-key-outcomes-from-cop26.

[7]. NASA Global Climate Change: Vital Signs of the Planet. "Global Surface Temperature," March 7, 2023. Accessed March 25, 2023. https://climate.nasa.gov/vital-signs/global-temperature/.
[8]. WMO Lead Centre Annual-to-Decadal Climate Prediction. "World Meteorological Organization." Accessed March 25, 2023. https://hadleyserver.metoffice.gov.uk/wmolc/.
[9]. Climate Action Tracker. "Temperatures," November 11, 2022. Accessed March 25, 2023. https://climateactiontracker.org/global/temperatures/.
[10]. Buis, Alan. "Why Milankovitch (Orbital) Cycles Can't Explain Earth's Current Warming." NASA Climate Change: Vital Signs of the Planet, February 27, 2022. Accessed March 25, 2023. https://climate.nasa.gov/ask-nasa-climate/2949/why-milankovitch-orbital-cycles-cant-explain-earths-current-warming.
[11]. Lindsey, Rebecca. "Climate Change: Atmospheric Carbon Dioxide." NOAA Climate.gov, June 23, 2022. Accessed March 25, 2023. https://www.climate.gov/news-features/understanding-climate/climate-change-atmospheric-carbon-dioxide.

[12]

. Mobius, Ariel. "How Do Greenhouse Gases Trap Heat in the Atmosphere?." MIT Climate Portal, February 19, 2021. Accessed March 25, 2023. https://climate.mit.edu/ask-mit/how-do-greenhouse-gases-trap-heat-atmosphere.
[13]. US EPA. "Overview of Greenhouse Gases," May 16, 2022. Accessed March 25, 2023. https://www.epa.gov/ghgemissions/overview-greenhouse-gases.
[14]. Scripps Institution of Oceanography. "The Keeling Curve." Accessed March 25, 2023. https://keelingcurve.ucsd.edu/.
[15]. US EPA. "Understanding Global Warming Potentials," May 5, 2022. Accessed March 25, 2023. https://www.epa.gov/ghgemissions/understanding-global-warming-potentials.
[16]. Wikipedia. "Global Warming Potential," March 18, 2023. Accessed March 25, 2023. https://en.wikipedia.org/wiki/Global_warming_potential.
[17]. US EPA. "Ozone-Depleting Substances," June 1, 2022. Accessed

March 25, 2023. https://www.epa.gov/ozone-layer-protection/ozone-depleting-substances.

[18]. Wikipedia. "Ozone Depletion," March 17, 2023. Accessed March 25, 2023. https://en.wikipedia.org/wiki/Ozone_depletion.

[19]. US EPA. "International Treaties and Cooperation about the Protection of the Stratospheric Ozone Layer," August 30, 2021. Accessed March 25, 2023. https://www.epa.gov/ozone-layer-protection/international-treaties-and-cooperation-about-protection-stratospheric-ozone.

[20]. UN News. "Ozone Layer Recovery Is on Track, Due to Success of Montreal Protocol," January 10, 2023. Accessed March 25, 2023. https://news.un.org/en/story/2023/01/1132277.

[21]. Franta, Benjamin. "On Its 100th Birthday in 1959, Edward Teller Warned the Oil Industry about Global Warming." The Guardian, January 1, 2018. Accessed March 25, 2023. https://www.theguardian.com/environment/climate-consensus-97-per-cent/2018/jan/01/on-its-hundredth-birthday-in-1959-edward-teller-warned-the-oil-industry-about-global-warming.; Supran, G., S. Rahmstorf, and N. Oreskes. "Assessing ExxonMobil's Global Warming Projections." *Science* 379, no. 6628 (January 13, 2023). https://doi.org/10.1126/science.abk0063.; Rannard, Georgina. "ExxonMobil: Oil Giant Predicted Climate Change in 1970s - Scientists." BBC News, January 12, 2023. Accessed March 25, 2023. https://www.bbc.com/news/science-environment-64241994.; Hall, Shannon. "Exxon Knew about Climate Change Almost 40 Years Ago." Scientific American, October 26, 2015. Accessed March 25, 2023. https://www.scientificamerican.com/article/exxon-knew-about-climate-change-almost-40-years-ago/.; Tabuchi, Hiroko. "Exxon Scientists Predicted Global Warming, Even as Company Cast Doubts, Study Finds." The New York Times, January 13, 2023. Accessed March 25, 2023. https://www.nytimes.com/2023/01/12/climate/exxon-mobil-global-warming-climate-change.html.; Pattee, Emma. "The Scientists Hired by Big Oil Who Predicted the Climate Crisis Long Ago." The Guardian, June 13, 2022. Accessed March 25, 2023. https://www.theguardian.com/environment/2021/jul/02/scientists-

climate-crisis-big-oil-climate-crimes.; Milman, Oliver. "Revealed: Exxon Made 'Breathtakingly' Accurate Climate Predictions in 1970s and 80s." The Guardian, January 16, 2023. Accessed March 25, 2023. https://www.theguardian.com/business/2023/jan/12/exxon-climate-change-global-warming-research.

[22]. Matthews, M. A. and Shell International Chemical Company Ltd. "The Earth's Carbon Cycle." *The New Scientist*, October 8, 1959, 644–46. Accessed March 25, 2023. https://www.documentcloud.org/documents/4447739-The-Earth-s-Carbon-Cycle-Shell.

[23]. Wikipedia. "Global Climate Coalition," February 4, 2023. Accessed March 25, 2023. https://en.wikipedia.org/wiki/Global_Climate_Coalition.; Yoder, Kate. "They Derailed Climate Action for a Decade. And Bragged about It." Grist, April 15, 2022. Accessed March 25, 2023. https://grist.org/accountability/how-the-global-climate-coalition-derailed-climate-action/.

[24]. Wikipedia. "Information Council for the Environment," January 4, 2022. Accessed March 25, 2023. https://en.wikipedia.org/wiki/Information_Council_for_the_Environment.

[25]. Union of Concerned Scientists. "The Disinformation Playbook," October 10, 2017. Accessed March 25, 2023. https://www.ucsusa.org/resources/disinformation-playbook.; Wikipedia. "Tobacco Industry Playbook," March 23, 2023. Accessed March 25, 2023. https://en.wikipedia.org/wiki/Tobacco_industry_playbook.

[26]. Wikipedia. "Climate Change Denial," March 16, 2023. Accessed March 25, 2023. https://en.wikipedia.org/wiki/Climate_change_denial.

[27]. OpenSecrets. "Oil & Gas Lobbying Profile." Accessed March 25, 2023. https://www.opensecrets.org/federal-lobbying/industries/summary?cycle=2022&id=E01.

[28]. OpenSecrets. "Oil & Gas." Accessed March 25, 2023. https://www.opensecrets.org/industries/indus.php?ind=E01.

[29]. Wikipedia. "Global Warming Controversy," March 1, 2023. Accessed March 25, 2023. https://en.wikipedia.org/wiki/

Global_warming_controversy.
[30]. Encyclopedia Britannica. "Industrial Revolution | Definition, History, Dates, Summary, & Facts," July 20, 1998. Accessed March 25, 2023. https://www.britannica.com/event/Industrial-Revolution.
[31]. Evans, Simon. "Solar, Wind and Nuclear Have 'Amazingly Low' Carbon Footprints, Study Finds." Carbon Brief, December 8, 2017. Accessed March 25, 2023. http://www.carbonbrief.org/solar-wind-nuclear-amazingly-low-carbon-footprints/.
[32]. Vaidyanathan, Gayathri and Climate Wire. "Political Football over Climate Change Rattles Windows of Ivory Tower." Scientific American, March 23, 2015. Accessed March 25, 2023. http://www.scientificamerican.com/article/political-football-over-climate-change-rattles-windows-of-ivory-tower/.
[33]. Laybourn, Laurie, Henry Throp, and Suzannah Sherman. "1.5°C – Dead or Alive? The Risks to Transformational Change from Reaching and Breaching the Paris Agreement Goal." Institute for Public Policy Research, February 16, 2023. Accessed March 25, 2023. https://www.ippr.org/research/publications/1-5c-dead-or-alive.

[34]

. US EPA. "Sources of Greenhouse Gas Emissions," August 5, 2022. Accessed March 25, 2023. http://www.epa.gov/ghgemissions/sources-greenhouse-gas-emissions.
[35]. US EPA. "Global Greenhouse Gas Emissions Data," February 15, 2023. Accessed March 25, 2023. http://www.epa.gov/ghgemissions/global-greenhouse-gas-emissions-data.
[36]. US EPA. "Sources of Greenhouse Gas Emissions," August 5, 2022. Accessed March 25, 2023. http://www.epa.gov/ghgemissions/sources-greenhouse-gas-emissions.
[37]. Roser, Max. "Why Did Renewables Become so Cheap so Fast?" Our World in Data, December 1, 2020. Accessed March 25, 2023. https://ourworldindata.org/cheap-renewables-growth.
[38]. Millennium Alliance for Humanity and the Biosphere (MAHB). "When Fossil Fuels Run Out, What Then?," May 23, 2019.

Accessed March 25, 2023. https://mahb.stanford.edu/library-item/fossil-fuels-run/.
[39]. Domonoske, Camila. "General Motors Sets All-Electric Target For Vehicles By 2035." NPR, February 1, 2021. Accessed March 25, 2023. http://www.npr.org/2021/02/01/962946561/general-motors-sets-all-electric-target-for-vehicles-by-2035.
[40]. Harvard T.H. Chan School of Public Health. "Home Is Where the Pipeline Ends," June 28, 2022. Accessed March 25, 2023. http://www.hsph.harvard.edu/c-change/news/home-is-where-the-pipeline-ends/.
[41]. Harvard T.H. Chan School of Public Health. "Natural Gas Used in Homes Contains Hazardous Air Pollutants," June 28, 2022. Accessed March 25, 2023. http://www.hsph.harvard.edu/c-change/news/natural-gas-used-in-homes/.
[42]. HowStuffWorks. "How Does a Halogen Light Bulb Work?," March 8, 2023. Accessed March 25, 2023. https://home.howstuffworks.com/question151.htm.
[43]. Harris, William. "How CFL Bulbs Work." HowStuffWorks, March 8, 2023. Accessed March 25, 2023. http://science.howstuffworks.com/environmental/green-tech/sustainable/cfl-bulb.htm.
[44]. Guzman, Francisco. "How LED Light Bulbs Work." HowStuffWorks, March 8, 2023. Accessed March 25, 2023. http://science.howstuffworks.com/environmental/green-tech/sustainable/led-light-bulb.htm.
[45]. Eartheasy. "LED Light Bulbs: Comparison Charts." Eartheasy Guides & Articles, February 15, 2022. Accessed March 25, 2023. https://learn.eartheasy.com/guides/led-light-bulbs-comparison-charts/.
[46]. US Department of Energy. "History of Air Conditioning," July 20, 2015. Accessed March 25, 2023. https://www.energy.gov/articles/history-air-conditioning.
[47]. US EPA. "Should You Have the Air Ducts in Your Home Cleaned?," February 27, 2023. Accessed March 25, 2023. https://www.epa.gov/indoor-air-quality-iaq/should-you-have-air-ducts-your-home-cleaned.

[48]

. US Department of Energy. "Ductless Mini-Split Air Conditioners." Accessed March 25, 2023. https://www.energy.gov/energysaver/ductless-mini-split-air-conditioners.

[49]. Baker, Thomas. "Read This Before You Buy Ductless AC." This Old House, March 13, 2023. Accessed March 25, 2023. https://www.thisoldhouse.com/heating-cooling/21018989/read-this-before-you-buy-ductless-ac.

[50]. US Energy Information Administration. "Efficiency Requirements for Residential Central AC and Heat Pumps to Rise in 2023," July 30, 2019. Accessed March 25, 2023. https://www.eia.gov/todayinenergy/detail.php?id=40232.

[51]. Energy Star. "ENERGY STAR Certified Heat Pumps (Mini-Split)." Accessed March 25, 2023. https://www.energystar.gov/productfinder/product/certified-mini-split-heat-pumps/.

[52]. Energy Star. "ENERGY STAR Certified Heat Pumps (Ducted)." Accessed March 25, 2023. https://www.energystar.gov/productfinder/product/certified-central-heat-pumps/.

[53]. Energy Star. "About ENERGY STAR." Accessed March 25, 2023. https://www.energystar.gov/.

[54]. Energy Star. "Kitchen Appliances." Accessed March 25, 2023. https://www.energy.gov/energysaver/kitchen-appliances.

[55]. Thompson, Dennis. "Gas Used in Homes Has Links to Cancer; Leaks Often Undetected." US News & World Report, July 1, 2022. Accessed March 25, 2023. https://www.usnews.com/news/health-news/articles/2022-07-01/gas-used-in-homes-has-links-to-cancer-leaks-often-undetected.; Michanowicz, Drew R., Archana Dayalu, Curtis L. Nordgaard, Jonathan J. Buonocore, Molly Wilson Fairchild, Robert C. Ackley, Jessica E Schiff, et al. "Home Is Where the Pipeline Ends: Characterization of Volatile Organic Compounds Present in Natural Gas at the Point of the Residential End User." *Environmental Science & Technology* 56, no. 14 (June 28, 2022): 10258–68. https://doi.org/10.1021/acs.est.1c08298.

[56]. Snyder, Robert. "Leukemia and Benzene." *International Journal of Environmental Research and Public Health* 9, no. 8 (August 14, 2012): 2875–93. https://doi.org/10.3390/ijerph9082875.
[57]. Jordan, Rob. "Rethinking Cooking with Gas." Stanford News, January 27, 2022. Accessed March 25, 2023. https://news.stanford.edu/2022/01/27/rethinking-cooking-gas/.
[58]. Dove, Laurie L. "Is Induction Cooking Better Than Gas or Electric?" HowStuffWorks, March 8, 2023. Accessed March 25, 2023. https://recipes.howstuffworks.com/tools-and-techniques/induction-cooking.htm?
srch_tag=3bvz23ntle7fhn4p74izqkdmqncjur2i.
[59]. US Department of Energy. "Tankless or Demand-Type Water Heaters." Accessed March 25, 2023. http://www.energy.gov/energysaver/tankless-or-demand-type-water-heaters.
[60]. US Department of Energy. "Laundry." Accessed March 25, 2023. https://www.energy.gov/energysaver/laundry.

[61]

. YouTube [TED]. "Shubhendu Sharma: How to Grow a Tiny Forest Anywhere," September 4, 2014. Accessed March 25, 2023. https://www.youtube.com/watch?v=3BgPFIKCaOQ.
[62]. US Department of Energy. "Weatherization Assistance Program." Accessed March 25, 2023. https://www.energy.gov/scep/wap/weatherization-assistance-program.
[63]. US Department of Energy. "Update or Replace Windows." Accessed March 25, 2023. https://www.energy.gov/energysaver/update-or-replace-windows.
[64]. US Department of Energy. "FOTW #1230, March 21, 2022: More than Half of All Daily Trips Were Less than Three Miles in 2021," March 21, 2022. Accessed March 25, 2023. https://www.energy.gov/eere/vehicles/articles/fotw-1230-march-21-2022-more-half-all-daily-trips-were-less-three-miles-2021.
[65]. Bustos, Laura Williams, M.S.. "What Is the Average Cycling Speed?" Bicycling, February 28, 2023. Accessed March 25, 2023. https://www.bicycling.com/training/a43031175/average-

cycling-speed/.

[66]. Ritchie, Hannah. "Which Form of Transport Has the Smallest Carbon Footprint?" Our World in Data, October 13, 2020. Accessed March 25, 2023. https://ourworldindata.org/travel-carbon-footprint.

[67]. Woodard, Collin. "This Electric Bus Drives 1,100 Miles on a Single Charge." MotorTrend, September 20, 2017. Accessed March 25, 2023. http://www.motortrend.com/news/electric-bus-claimed-1100-mile-range/.

[68]. US EPA. "Plug-in Electric Vehicle Charging," February 22, 2023. Accessed March 25, 2023. http://www.epa.gov/greenvehicles/plug-electric-vehicle-charging.

[69]. US Department of Energy. "Alternative Fuels Data Center: Charging Electric Vehicles at Home." Accessed March 25, 2023. https://afdc.energy.gov/fuels/electricity_charging_home.html.

[70]. US Department of Energy. "FOTW #1190, June 14, 2021: Battery-Electric Vehicles Have Lower Scheduled Maintenance Costs than Other Light-Duty Vehicles," June 14, 2021. Accessed March 25, 2023. http://www.energy.gov/eere/vehicles/articles/fotw-1190-june-14-2021-battery-electric-vehicles-have-lower-scheduled.

[71]. Sivak, Michael, and Brandon Schoettle. "Relative Costs of Driving Electric and Gasoline Vehicles in the Individual U.S. States." National Academies: Transportation Research Board, January 2018. Accessed March 31, 2023. https://trid.trb.org/view/1508116.

[72]. Union of Concerned Scientists. "Cleaner Cars from Cradle to Grave," October 29, 2015. Accessed March 25, 2023. https://www.ucsusa.org/resources/cleaner-cars-cradle-grave.

[73]. US Department of Energy. "All-Electric Vehicles." Accessed March 25, 2023. https://fueleconomy.gov/feg/evtech.shtml.

[74]

. US Department of Energy. "2023 Chevrolet Bolt EV." Accessed March 25, 2023. https://www.fueleconomy.gov/feg/Find.do?action=sbs&id=45751.

[75]. Go Climate. "Travel Emissions Calculator." Accessed March 25, 2023. https://www.goclimate.com/us/travel-calculator.
[76]. Ritchie, Hannah. "Which Form of Transport Has the Smallest Carbon Footprint?" Our World in Data, October 13, 2020. Accessed March 25, 2023. https://ourworldindata.org/travel-carbon-footprint.
[77]. Bryan, Victoria. "Hawaiian Airlines Invests in Electric Seaglider Maker for Inter-Island Flights - AeroTime." AeroTime, May 11, 2022. Accessed March 25, 2023. http://www.aerotime.aero/articles/30983-hawaiian-airlines-electric-seaglider-regent.; Associated Press. "Hawaiian Airlines Exploring Electric Airplane Technology." US News & World Report, May 12, 2022. Accessed March 25, 2023. https://www.usnews.com/news/best-states/hawaii/articles/2022-05-12/hawaiian-airlines-exploring-electric-airplane-technology.
[78]. Archer Aviation. "United Airlines And Archer Announce First Commercial Electric Air Taxi Route In Chicago," March 23, 2023. Accessed March 25, 2023. http://www.archer.com/news/united-airlines-and-archer-announce-first-commercial-electric-air-taxi-route-in-chicago.
[79]. Thomas, Ian. "United Airlines Is Aiming to Have Electric Planes Flying by 2030." CNBC, October 10, 2022. Accessed March 25, 2023. http://www.cnbc.com/2022/10/06/united-airlines-is-aiming-to-have-electric-planes-flying-by-2030.html.
[80]. Airbus. "Hybrid and Electric Flight." Accessed March 27, 2023. http://www.airbus.com/en/innovation/zero-emission-journey/hybrid-and-electric-flight.
[81]. Maron, Dina Fine. "WHO Moves to Contain Superbugs on the Farm." Scientific American, November 7, 2017. Accessed March 25, 2023. http://www.scientificamerican.com/article/who-moves-to-contain-superbugs-on-the-farm/.
[82]. "The State of Food Security and Nutrition in the World 2022." *FAO EBooks*, July 6, 2022. https://doi.org/10.4060/cc0639en.
[83]. Batten, Robert, Jennifer L Turner, and Susan Chan Shifflett. "China's Upstream Advantage in the Great Himalayan Watershed - The National Bureau of Asian Research (NBR)." The National

Bureau of Asian Research (NBR), July 15, 2013. Accessed March 25, 2023. http://www.nbr.org/publication/chinas-upstream-advantage-in-the-great-himalayan-watershed/.

[84]. Gutierrez, Miren, and Guy Jobbins. "China's Distant-Water Fishing Fleet: Scale, Impact and Governance." ODI, June 2, 2020. Accessed March 25, 2023. http://odi.org/en/publications/chinas-distant-water-fishing-fleet-scale-impact-and-governance/.

[85]

. Trollman, Hana. "Here's How the Russian Invasion Has Impacted Ukraine Crop Exports." World Economic Forum, April 20, 2022. Accessed March 25, 2023. http://www.weforum.org/agenda/2022/04/ukraine-grain-shortages-russia-conflict/.

[86]. Polakovic, Gary. "Water Dispute on the Nile River Could Destabilize the Region." USC News, July 13, 2021. Accessed March 25, 2023. http://news.usc.edu/188414/nile-river-water-dispute-filling-dam-egypt-ethiopia-usc-study/.

[87]. Sutherland, Paige. "The Water Fight over the Shrinking Colorado River." BBC News, April 8, 2021. Accessed March 25, 2023. https://www.bbc.com/news/world-us-canada-56608180.

[88]. USDA Economic Research Service. "U.S. Food Imports," March 23, 2023. Accessed March 25, 2023. https://www.ers.usda.gov/data-products/u-s-food-imports/.

[89]. Project Drawdown. "Table of Solutions," August 16, 2021. Accessed March 25, 2023. https://www.drawdown.org/solutions/table-of-solutions.

[90]. Hawkin, Paul. "Drawdown." Resilience, June 21, 2017. Accessed March 25, 2023. https://www.resilience.org/resources/drawdown/.

[91]. Springmann, Marco, H. Charles J. Godfray, Mike Rayner, and Peter Scarborough. "Analysis and Valuation of the Health and Climate Change Cobenefits of Dietary Change." *Proceedings of the National Academy of Sciences of the United States of America* 113, no. 15 (April 12, 2016): 4146–51. https://doi.org/10.1073/pnas.1523119113.

[92]. Kerlin, Kat. "The Diet That Helps Fight Climate Change." UC

Davis, December 28, 2017. Accessed March 25, 2023. https://www.ucdavis.edu/news/diet-helps-fight-climate-change.
[93]. Carrington, Damian. "Avoiding Meat and Dairy Is 'Single Biggest Way' to Reduce Your Impact on Earth." The Guardian, May 31, 2018. Accessed March 25, 2023. http://www.theguardian.com/environment/2018/may/31/avoiding-meat-and-dairy-is-single-biggest-way-to-reduce-your-impact-on-earth.; Poore, Joseph, and Thomas Nemecek. "Reducing Food's Environmental Impacts through Producers and Consumers." *Science* 360, no. 6392 (June 1, 2018): 987–92. https://doi.org/10.1126/science.aaq0216.
[94]. Ladyzhets, Betsy. "How Much Does Eating Meat Affect Nations' Greenhouse Gas Emissions?" Science News, May 5, 2022. Accessed March 25, 2023. http://www.sciencenews.org/article/food-emissions-data-diet-carbon-greenhouse-gas-climate-agriculture.
[95]. Peters, Christian J., Jamie Picardy, Amelia Darrouzet-Nardi, Jennifer L. Wilkins, Timothy J. Griffin, and Gary W. Fick. "Carrying Capacity of U.S. Agricultural Land: Ten Diet Scenarios." *Elementa* 4 (July 22, 2016). https://doi.org/10.12952/journal.elementa.000116.
[96]. New England Vegetable Guide. "Table 15: Approximate Yields," May 26, 2022. Accessed March 25, 2023. http://nevegetable.org/cultural-practices/table-15-approximate-yields.
[97]. Sas, Wojciech. "Cattle per Acre Calculator." OMNI Calculator, January 18, 2023. Accessed March 25, 2023. https://www.omnicalculator.com/biology/cattle-per-acre.

[98]

. Carrington, Damian. "Avoiding Meat and Dairy Is 'Single Biggest Way' to Reduce Your Impact on Earth." The Guardian, May 31, 2018. Accessed March 25, 2023. http://www.theguardian.com/environment/2018/may/31/avoiding-meat-and-dairy-is-single-biggest-way-to-reduce-your-impact-on-earth.
[99]. Bland, Alastair. "Is the Livestock Industry Destroying the Planet?" Smithsonian Magazine, August 1, 2012. Accessed March

25, 2023. http://www.smithsonianmag.com/travel/is-the-livestock-industry-destroying-the-planet-11308007/.

[100]. Ritchie, Hannah. "Half of the World's Habitable Land Is Used for Agriculture." Our World in Data, November 11, 2019. Accessed March 25, 2023. https://ourworldindata.org/global-land-for-agriculture.

[101]. Ritchie, Hannah, and Max Roser. "Water Use and Stress." Our World in Data, July 2018. Accessed March 25, 2023. http://ourworldindata.org/water-use-stress#share-of-freshwater-withdrawals-used-in-agriculture.

[102]. Water Footprint of Food Guide - Water Footprint Calculator. (2022, November 23). *Water Footprint Calculator*. https://www.watercalculator.org/water-footprint-of-food-guide/

[103]. Water Footprint Calculator. "How Much Water Does It Take to Make a Hamburger?," July 15, 2022. Accessed March 25, 2023. http://www.watercalculator.org/footprint/what-is-the-water-footprint-of/.

[104]. Worldometer. "World Population Clock: 8 Billion People (LIVE, 2023)." Accessed March 25, 2023. https://www.worldometers.info/world-population/.

[105]. Carrington, Damian. "Humans Just 0.01% of All Life but Have Destroyed 83% of Wild Mammals – Study." The Guardian, May 21, 2018. Accessed March 25, 2023. http://www.theguardian.com/environment/2018/may/21/human-race-just-001-of-all-life-but-has-destroyed-over-80-of-wild-mammals-study.

[106]. True Cost. "List of Foods By Environmental Impact and Energy Efficiency," January 26, 2016. Accessed March 25, 2023. http://truecostblog.com/2010/02/24/list-of-foods-by-environmental-impact-and-energy-efficiency/.

[107]. World Health Organization. "Healthy Diet," April 29, 2020. Accessed March 25, 2023. https://www.who.int/news-room/fact-sheets/detail/healthy-diet.

[108]. Centers for Disease Control and Prevention (CDC). "Poor Nutrition." Accessed March 25, 2023. http://www.cdc.gov/chronicdisease/resources/publications/factsheets/nutrition.htm.

[109]. National Institutes of Health (NIH). "How Dietary Factors

Influence Disease Risk," March 14, 2017. Accessed March 25, 2023. https://www.nih.gov/news-events/nih-research-matters/how-dietary-factors-influence-disease-risk.

[110]

. Campbell, T. Colin, & Thomas M. Campbell. *The China Study: The Most Comprehensive Study of Nutrition Ever Conducted and the Startling Implications for Diet, Weight Loss and Long-Term Health.* BenBella Books, 2006.

[111]. Esselstyn, Caldwell B. "Updating a 12-Year Experience with Arrest and Reversal Therapy for Coronary Heart Disease (an Overdue Requiem for Palliative Cardiology)." *American Journal of Cardiology* 84, no. 3 (August 1, 1999): 339–41. https://doi.org/10.1016/s0002-9149(99)00290-8.

[112]. O'Keefe, Sean F., Jia Li, Leo Lahti, Junhai Ou, Franck Carbonero, Khaled Mohammed, Joram M. Posma, et al. "Fat, Fibre and Cancer Risk in African Americans and Rural Africans." *Nature Communications* 6, no. 1 (April 28, 2015). https://doi.org/10.1038/ncomms7342.

[113]. Blue Zones. "History of Blue Zones," April 30, 2021. Accessed March 25, 2023. https://www.bluezones.com/about/history/.

[114]. Cleveland Clinic. "What Are Complete Proteins?," December 6, 2022. Accessed March 25, 2023. http://health.clevelandclinic.org/do-i-need-to-worry-about-eating-complete-proteins/.

[115]. Wikipedia. "Global Industry Classification Standard," March 21, 2023. Accessed March 25, 2023. http://en.wikipedia.org/wiki/Global_Industry_Classification_Standard.

[116]. McFall-Johnsen, Morgan. "These Facts Show How Unsustainable the Fashion Industry Is." World Economic Forum, January 31, 2020. Accessed March 25, 2023. http://www.weforum.org/agenda/2020/01/fashion-industry-carbon-unsustainable-environment-pollution/.

[117]. OEcotextiles. "What Is the Energy Profile of the Textile Industry?," October 31, 2011. Accessed March 25, 2023. http://oecotextiles.blog/2009/06/16/what-is-the-energy-profile-of-the-textile-industry/.; Schlesinger, William. "The

Fabric for Our Lives." Duke University: Nicholas School of the Environment, November 26, 2018. Accessed March 25, 2023. http://blogs.nicholas.duke.edu/citizenscientist/the-fabric-for-our-lives/.; Barrett, John, Matthew Chadwick, and Mike Chadwick. "Ecological Footprint and Water Analysis of Cotton, Hemp and Polyester." Stockholm Environment Institute, March 11, 2020. Accessed March 25, 2023. http://www.sei.org/publications/ecological-footprint-water-analysis-cotton-hemp-polyester/.

[118]. Textile Exchange. "Polyester," January 3, 2023. Accessed March 25, 2023. https://textileexchange.org/polyester/.

[119]. Drew, Deborah, and Genevieve Yehounme. "The Apparel Industry's Environmental Impact in 6 Graphics." World Resources Institute, July 5, 2017. Accessed March 25, 2023. https://www.wri.org/insights/apparel-industrys-environmental-impact-6-graphics.

[120]. Ellen MacArther Foundation. "Fashion and the Circular Economy." Accessed March 25, 2023. https://ellenmacarthurfoundation.org/fashion-and-the-circular-economy-deep-dive.

[121]. McFall-Johnsen, Morgan. "These Facts Show How Unsustainable the Fashion Industry Is." World Economic Forum, January 31, 2020. Accessed March 25, 2023. https://www.weforum.org/agenda/2020/01/fashion-industry-carbon-unsustainable-environment-pollution/.

[122]. Hayes, Adam. "Fast Fashion Explained and How It Impacts Retail Manufacturing." Investopedia, September 16, 2022. Accessed March 25, 2023. https://www.investopedia.com/terms/f/fast-fashion.asp.

[123]. McFall-Johnsen, Morgan. "These Facts Show How Unsustainable the Fashion Industry Is." World Economic Forum, January 31, 2020. Accessed March 25, 2023. https://www.weforum.org/agenda/2020/01/fashion-industry-carbon-unsustainable-environment-pollution/.

[124]. US EPA. "Textiles: Material-Specific Data," December 3, 2022. Accessed March 25,

2023. https://www.epa.gov/facts-and-figures-about-materials-waste-and-recycling/textiles-material-specific-data.
[125]. McFall-Johnsen, Morgan. "These Facts Show How Unsustainable the Fashion Industry Is." World Economic Forum, January 31, 2020. Accessed March 25, 2023. https://www.weforum.org/agenda/2020/01/fashion-industry-carbon-unsustainable-environment-pollution/.
[126]. Williams, Alan. "Washing Clothes Releases Thousands of Microplastic Particles into Environment, Study Shows." University of Plymouth, September 27, 2016. Accessed March 25, 2023. https://www.plymouth.ac.uk/news/washing-clothes-releases-thousands-of-microplastic-particles-into-environment-study-shows.; Napper, Imogen E., and Richard F. Thompson. "Release of Synthetic Microplastic Plastic Fibres from Domestic Washing Machines: Effects of Fabric Type and Washing Conditions." *Marine Pollution Bulletin* 112, no. 1–2 (November 15, 2016): 39–45. https://doi.org/10.1016/j.marpolbul.2016.09.025.
[127]. Phys.org. "Microfibers in the Mediterranean Sea Are Floating Homes for Bacteria," November 30, 2022. Accessed March 25, 2023. https://phys.org/news/2022-11-microfibers-mediterranean-sea-homes-bacteria.html.; Pedrotti, Maria Luiza, Ana Luzia De Figueiredo Lacerda, Stephanie Petit, Jean François Ghiglione, and Gabriel Gorsky. "Vibrio Spp and Other Potential Pathogenic Bacteria Associated to Microfibers in the North-Western Mediterranean Sea." *PLOS ONE* 17, no. 11 (November 30, 2022): e0275284. https://doi.org/10.1371/journal.pone.0275284.
[128]. Torchinsky, Rina. "For the First Time, Researchers Find Microplastics Deep in the Lungs of Living People." NPR, April 6, 2022. Accessed March 25, 2023. https://www.npr.org/2022/04/06/1091246691/microplastics-found-in-human-lungs.
[129]. Carrington, Damian. "Microplastics Found in Human Breast Milk for the First Time." The Guardian, October 7, 2022. Accessed March 25, 2023. https://www.theguardian.com/environment/2022/oct/07/microplastics-human-breast-milk-

first-time.

[130]. Ragusa, Antonio, Alessandro Svelato, Criselda Santacroce, Piera Catalano, Simona Sabbatini, Oliana Carnevali, Fabrizio Papa, et al. "Plasticenta: First Evidence of Microplastics in Human Placenta." *Environment International* 146 (January 1, 2021): 106274. https://doi.org/10.1016/j.envint.2020.106274.

[131]

. Leslie, H. M., Martin Van Velzen, Sicco H. Brandsma, Dick Vethaak, Juan J. Garcia-Vallejo, & Marja H. Lamoree. "Discovery and Quantification of Plastic Particle Pollution in Human Blood." *Environment International* 163 (March 1, 2022): 107199. https://doi.org/10.1016/j.envint.2022.107199.

[132]. Carrington, Damian. "Bottle-Fed Babies Swallow Millions of Microplastics a Day, Study Finds." The Guardian, October 20, 2020. Accessed March 25, 2023. https://www.theguardian.com/environment/2020/oct/19/bottle-fed-babies-swallow-millions-microplastics-day-study.; Li, Dunzhu, Yunhong Shi, Luming Yang, Liwen Xiao, Daniel K. Kehoe, Yurii K. Gun'ko, John J. Boland, and Jing Wang. "Microplastic Release from the Degradation of Polypropylene Feeding Bottles during Infant Formula Preparation." *Nature Food* 1, no. 11 (November 1, 2020): 746–54. https://doi.org/10.1038/s43016-020-00171-y.

[133]. Ali, Shirin. "Massive Mountain of Clothing in the Desert Is Growing Even Bigger." The Hill, November 8, 2021. Accessed March 25, 2023. https://thehill.com/changing-america/sustainability/environment/580521-massive-mountain-of-clothing-in-the-desert-is/.

Made in the USA
Monee, IL
15 July 2026

57281840R00066